Architecture Follows Fish

The United States Court at the Fisheries Exhibition, London, 1883.
Illustration by Frederick Whymper, *The Fisheries of the World*, 1894.

Architecture Follows Fish
An Amphibious History of the North Atlantic

André Tavares

THE MIT PRESS
Cambridge, Massachusetts . London, England

Myre freezing plant was a concrete mass with the whitewashed exterior typical of a modern architectural monument. Inaugurated in 1955 in Norway's Vesterålen archipelago, it has been praised as "a sober yet optimistic and handsome piece of industrial architecture"[1] with functional asymmetries from the repeating wide square windows to the extensions and entrances that were soon added. Once it began operations, Arctic cod populations migrating from the Barents Sea southward through the Egga ridge in search of capelin or heading for their spawning grounds met a structure purpose-built for processing them into frozen fillets.

The machine-like elegance of the Myre freezing plant, so different from the preexisting timber constructions of Northern Norway's fishing villages, epitomizes the postwar transformation of the region's economy and landscape.[2] Until the middle of the twentieth century, the fishing villages of Vesterålen, like those of Finnmark and the Lofoten archipelago, were distinguished by building types related to the steps of fish processing: docks, sheds for fishing gear, structures for processing and storing the catch, and houses for fisherfolk. In such settlements, cod was transformed into *støckfisk*, *tørrfisk*, or *klippfisk* by means of salting and drying. Spectacular wooden drying racks dispersed through a mountainous and jagged coastal topography marked stretches of human occupation. Thus, the concrete volume of the freezing plant, which vertically concentrated the functions inherited from previous fishing practices in a single building,[3] was a definitive formal departure from tradition. The construction of the plant also reflects the vertical integration of production into a cooperative organization, a strategy promoted by Norway's social welfare policy. Above all, the freezer introduced an expensive high-tech facility to Myre that could be used on a large scale. Refrigeration redirected the entire dynamic of fisheries, changing everything from food consumption habits to the pressures exerted upon marine ecosystems, and the architecture of refrigeration embodies this socioecological chain of events.

Fish processing is adapted to fish physiology, so it follows that fishing architecture varies according to the species targeted. A telling comparison is between cod and herring, which produce very different architectures. Cod is a large demersal fish with an elevated

Vesterålen archipelago,
Myre, ca. 1950. Photo:
Fjellanger Wideraøe.
Courtesy Museum Nord,
Øksnes Museum.

position in the trophic chain: it preys on species such as capelin or herring. Herring, on the other hand, is a pelagic fodder fish, living in the upper levels of the water column and feeding on plankton. Predator and prey fish have different physiologies: cod is large, muscular, and low in fat; herring is smaller and rich in fatty acids. Cod keeps well when salted and dried; herring is good for smoking. The latter requires a smoking house such as the one depicted in a well-known cross section by Duhamel du Monceau (1700–1782).[4] In it, we see a chimneyless brick structure about 14 meters tall to the ridge of the roof that is divided into three equal sections of about 3 meters in width. Around a fire in the center of the building, each of the modules is subdivided by ladders from which acrobatic workers organize rows of herring sticks. Filled with herring and smoke, the smokehouse transforms the fish into a product ready to be packed in barrels and distributed for consumption. Du Monceau's engraving—a meticulous description that draws on the visual language of the *Encyclopédie*[5]—emphasizes the operation's tectonics, with bricks and herring shown as constitutive elements of the manufacturing process.[6] There is no apparent difference between brick and fish, between geological and biological elements, as if the extraordinary structure was actually built of herring.

Herring smokehouses and cod-freezing factories are two architectural forms corresponding to different fish-processing practices. But before they reach such buildings, fish must be caught. In a memorable scene in *Stromboli*, a classic neorealist film directed by Roberto Rossellini (1906–1977), the beautiful Karin, played by Ingrid Bergman (1915–1982), is horrified by the massacre of a tuna shoal in a fish trap in the Mediterranean Sea.[7] Similar devices were used by Iberian tuna fisheries in the Atlantic and set up during the season when the migratory tuna swim close to shore. Known as *almadrabas*, they are made up of nets suspended from buoys and anchored to the sea bottom with a central enclosure that can be as large as 150 by 50 meters. Fish are directed to the center by arms extending up to 4.5 kilometers toward the shore in one direction and out to sea in the other. Once the trap is full of prey, fishers encircle it, close the entrance, and haul the nets to the surface. With the help of spikes, they pull the large, heavy tuna, one by one, into boats. The operation begins somewhat calmly, but soon the imprisoned tuna become agitated and violence takes over: dripping blood mixes with saltwater, the captured fish flap frantically on the decks, and fishers jump into the sea to help with the catch. Italians call the operation *mattanza*, a massacre.

Fish traps are designed in accordance with the behavior and movements of the animals and territorialize the waters, establishing an intricate relationship between land and sea. The architectural form of the *almadrabas* off the southern coasts of the Iberian peninsula follows the instinctual trajectory of the tuna back and forth between the Atlantic and their spawning grounds in the Mediterranean. These two routes define the positioning of the trap arms, their specific location in the sea, and their relation to shore. In the nineteenth century, a Portuguese fishing expert suggested that the state should regulate the placement of the traps to ensure that the gear set for the inbound catch in May would not obstruct the outbound fish in June.[8] Such a legislative argument that seeks to maximize the extraction of natural resources reveals the connection between fish behavior and building practices at sea, and emphasizes just how much the design of the *almadrabas* combines biological and architectural knowledge.

It is common practice to associate these fishing architectures with vernacular building techniques, describing constructions that are deeply rooted in a place in connection with environmental constraints and local culture. Yet, as this book demonstrates, the history is more complicated than that. The North Atlantic is the habitat of many species, and beyond any cultural history of architecture there is an ecological history. Many building practices are borrowed from distant waters and shaped by unexpected dynamics. Hence, the following chapters weave together examples from many locations around the ocean, considering various local histories to foster a pan-Atlantic perspective on cultural, technological, and ecological exchanges. They parallel Newfoundland's history with that of European coastal communities and relate the dynamics of Norwegian ecosystems to American technologies, French coastal policies, and the development of British railways. Although the route seems tortuous at times, the geographical trajectory of the chapters aims to supplant the atavistic notion of the vernacular with the concept of architecture as belonging to multiple socioecological systems.

Contemporary fishing architecture conceals the relationship between building practices and the pressure they put on ecological systems, a relationship that is much more evident in the remnants of fishing architecture from around the turn of the twentieth century because the mechanisms connecting social activities to their ecological environment are easier to see. For this reason, lessons we find in the history of fishing architecture are important in tackling some of today's critical challenges, the first of which is to be able to consider the full range of ecological impacts associated with the transformation of the built environment. Second, they lead to an understanding of how architecture exists at the intersection of transformations in technologies and a much wider network of interests that range from marine ecology to cultural dimensions as intimate as eating habits. And third, they contribute to a global history, showing how deeply interrelated the various coasts of the North Atlantic are by seeing how fish can establish architectural continuities across a large body of water.

Because this book addresses, to use Bernard Rudofsky's term, "non-pedigreed" architecture,[9] the constructions we will encounter are often rather banal. Additionally, their visible ecological impact carries a destructive pathos that prevents us from being seduced into a nostalgic view of heritage. Yet, in their banality and perversity, many of these structures are just as spectacular as more acclaimed forms of architecture, creating innovative spaces and environments and providing new atmospheres in which to experience both landscapes and buildings. Such human inventiveness

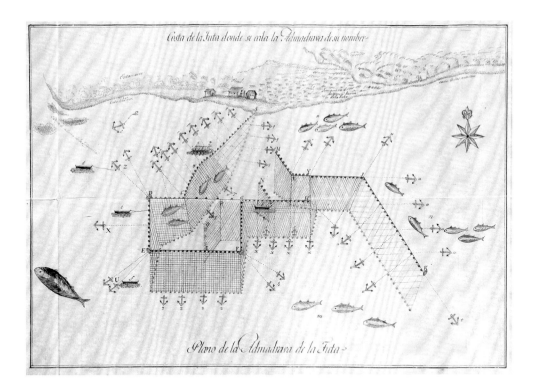

Costa de la Tuta donde se cala la Almadrava de su nombre

Plano de la Almadrava de la Tuta

results from specific types of fish, demonstrating how built shapes can result from cross-species entanglements.

To dissect and seek to understand such relationships between fish and architecture, the conceptual structure of this book layers five analytical perspectives. The first is biological: as we have seen, the physiological characteristics of fish trigger most of the architectural solutions. The second perspective is technological and considers fishing gear and navigation techniques. The third relates to fish processing and how the terrestrial activities associated with fish justify specific built apparatus. The fourth is political, as significant social spheres are involved in decisions that have a major impact both on land and at sea. And, finally, the fifth perspective analyzes consumption to associate human eating habits with the pressure they put upon ecosystems. These layers are interconnected: the characteristics and transformations of each one depend upon the others and influence them in turn. Biology, technology, processing, politics, and consumption are the keys to understanding the extent to which a fish shapes landscapes and the architecture to be found in them.

Tuna *almadraba*, La Tuta, 1831. Courtesy Archivo General Fundación Casa Medina Sidonia, Cádiz.

5

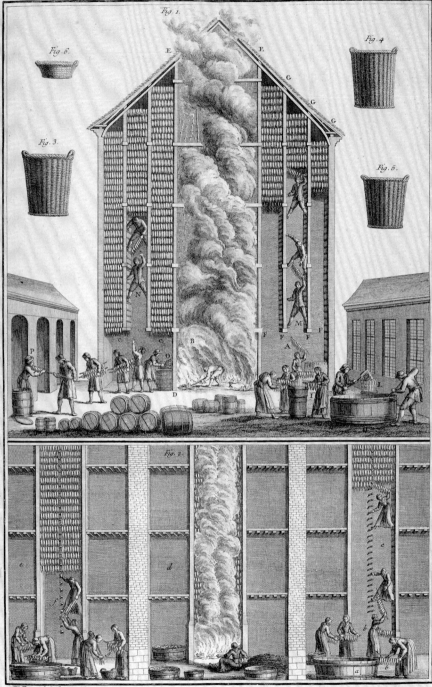

The historical period under examination runs from the Industrial Revolution to the mid-twentieth century when the Law of the Seas was implemented, the cold chain expanded, and consumption of frozen fillets grew. The globalization of markets at the end of the twentieth century introduced a new era of fisheries: current consumption is at unprecedented levels, illegal fisheries use excessively violent and cruel techniques,[10] and the ocean's biodiversity is compromised by an ever-growing list of endangered species. In parallel, aquaculture and fish farming have increased exponentially in terms of both production and consumption. To a certain extent, the very biology of fish is being transformed. Contemporary fishing architecture is preoccupied with aquaculture and, as in the wild fisheries, the primary product has generated a powerful architectural counterpart. But this is where this book ends.

facing Herring preparation and smokehouse, in Duhamel du Monceau, *Traité général des pêches et histoire des poissons*, vol. 2, section 3, plate 15, 1772. Courtesy Bibliothèque Nationale de France.

To what extent can fish produce architecture? It is obvious that fish do not build buildings, but their biological traits do generate architecture. Or at least this is the hypothesis that I want to explore—and demonstrate—in this book.[11] Fish species have a physiology and inhabit dynamic ecological systems where predation, after reaching a certain scale, gives form to particular architectures and landscapes. Once built, the architectures supporting fisheries have an impact on the ecological balance of fish populations. One can argue that there is a cod architecture, just as there is a sardine architecture, a sole architecture, or a hake architecture. There are as many architectures as there are ways to fish, which could suggest that such construction practices result from the fisheries and not from the fish. This might be a legitimate quibble, but it is worth focusing on fish to place our perception of the built phenomena solidly in the framework of the ecological transformation of the planet, an important alternative to the more common analysis of architecture as a product of human social practices. When we flip the lens of marine biology toward fish species and their behaviors to see how these prompt transformations on land—and also look at marine ecosystems in terms of terrestrial building processes—we realize that architecture has a history connected with the animals in the water, the dynamics of the oceans, and with marine biology.

7

Pedrouços fishing harbor, Lisbon, 1971. Photo: Gustavo Leitão. Courtesy Arquivo dos Portos de Lisboa, Setúbal e Sesimbra, APLSS/A8/19/1080.

Significant differences in the landscapes of cod and sardine fishing are rooted in differences in the physiology, the behavior, and the habitats of these species. The low-fat muscular flesh of cod lends itself to being dried, preserving its nutritional qualities for a long period, and so a cod-fishing landscape includes drying racks spread over large areas of windy terrain. By contrast, sardine, rich in fatty acids and subject to quick putrefaction after leaving the water, is better suited to canning in factories with quick access to fishing grounds and connections to logistical infrastructures. The form of each fishing landscape depends on the relationship between terrestrial and marine ecosystems. The relationship is not deterministic—to a fish, there is not necessarily a landscape—and there are plenty of contributing factors, be they social, political, or technical, that have made cod and sardine landscapes so different. Moreover, the architecture relates to each species within a precise historical moment; it is dependent on the circumstances of the growth of a species, and also of its decline and abandonment. Nonetheless, one can look at structures built on land and recognize the way they exploit, benefit from, and intrude upon fish and the larger marine ecosystems. But even more importantly, we should look to the sea and recognize the impact of terrestrial infrastructure there.

Fishing architecture exists at the intersection of various technologies and must accommodate fishing gear and food-processing

8

machinery, navigational instruments, and commercial logistics. Following the fish through the process of its transformation, we find extraordinary constructions designed to support trade and consumption, including drying racks, cold storage and warehouses, piers and moles, harbor schooners, steamers and other maritime vessels, and depots and market halls. Fishing architecture ranges from precarious constructions that accompany the seasonal life cycles of different species across coastal landscapes to factories that process enormous quantities of fish in controlled environments. Nets are another form of fishing architecture, adapted to the particular species they target and used or installed in the sea in response to fish behaviors: these are often large structures that respond to oceanographic conditions such as currents and water temperature. Fish traps, nets, and buoys territorialize marine landscapes by lending their names to places inconspicuous beyond the horizon line.

These built devices establish places that become part of navigational routes and lines on maps traced by vessels. Such routes connect land and sea in a continuum between coastal locations and fishing grounds, implying movement and reciprocity between marine ecosystems and land constructions. My interest in this architecture resides not in the shape of the buildings or in the virtues of the architects but in the desire to understand built manifestations of the human presence within nature. Fishing has an impact on the fragile equilibrium of ecosystems; humans are voracious predators and create objects that scale up their exploitation and extraction of natural resources. The architecture of fishing reveals multiple aspects of this permanent interaction between land and sea, demonstrating that the sea is a place and not a mere line on the horizon. What we build on land has consequences for marine ecosystems.

Looking at fish as propellers of the built environment allows us to overcome anthropocentric and nationalistic biases and consider the architecture of fisheries within the context of the environmental dynamics of the North Atlantic. The impact of fish on cities and landscapes is often unperceived and unacknowledged in architectural and urban narratives, but fish behavior and characteristics have made certain urban expansions possible as well as, on occasion, setting limits on the transformation of landscapes. This ecological dimension is a key agent that acts on the built environment and by extension on the marine ecosystems transformed by building practices on land. Fish are a powerful connector: considering architectures designed to accommodate fisheries around the ocean lays bare how constructions often seen as isolated phenomena due to the nationalistic bias of history; the celebrated local identities evident in many studies of vernacular architectures are actually shaped by cultural and technological transfers that create similar patterns in different places. Although our seas today are more territorialized than ever, with borders and Exclusive Economic Zones (EEZ), fish still have no nationality. There is no Icelandic cod or Portuguese sardine, since each species lives in complex geographical distributions, and their behavior is a response to the conditions of marine ecosystems such as water temperature, currents, nutrient abundance, and predators. Cod fished in Iceland share their genetic identity with cod from the eastern coast of Greenland;[12] the schools of sardines off the coast of Portugal have the same genetic characteristics as the sardines that regularly move between Morocco and the Bay of Biscay.[13] Such autonomy from political borders collides with fishing histories, usually conceived according to national, cultural, and social circumstances. Contemporary historians are making efforts to overcome the limitations of national histories to achieve an integrated understanding of the ecological dynamics of the ocean, shifting the narrative from human feats of ingenuity to the role of humans within ecosystems and our impact on biodiversity.[14] By tracing an ecological history of architecture—a history that sees built artifacts as part of complex ecosystems and not just as the product of human action—we get an image that collides with the narratives of territorial formation and transformation.

Cod was the starting point for this research. Mark Kurlansky considers it "the fish that changed the world," and recent research projects have underlined how in the fourteenth century its harvesting, processing, and consumption contributed to the Renaissance and subsequent cultural transformations along the shores of the Atlantic and throughout the Western world.[15] Today, cod is mostly sold in frozen form, with fresh fish available in times of abundance, but cured cod is still produced, often in peculiar forms. After being captured in Norwegian or Icelandic waters, the cod is landed and gutted for immediate freezing; it is then transported

to a processing unit (close to the landing harbors or as far away as Portugal and China), where it is defrosted and dressed. Next, it is cured in salt for a few months. The salting is followed by drying, achieved through mechanical ventilation. Once dry, some of the production is commercialized, while the rest, in ever-increasing amounts, moves to a new processing stage. The dried fish is cut and rehydrated (and the high-grade quota is deboned) to be deep-frozen, packed, and sold as a ready-to-cook product. This means that, after its capture, the cod follows a long trajectory of processing and commercialization.

The most time-sensitive, yet crucial, stage of the process for cod and for other species is dressing the fish. This is when the taste and nutritional qualities of the product are defined. When it comes to fresh fish, this stage can be delayed until the animal is cooked, generating lower processing costs and higher product value. In other forms of processing, such as salting, pickling, and smoking, dressing is the key for the success of the further

Fish being landed from the trawler at the auction market, Lisbon, ca. 1966. Biblioteca Central de Marinha/Arquivo Histórico, FG/009/11/010/004.

Fisherfolk's sheds, The
Stade, Hastings, 1956.
Photo: Eric de Mare.
Courtesy Architectural
Press Archive, RIBA
Collections 17258.

steps and the preservation of the fish. The remnants, consisting
of heads, guts, and even bones, are often used to produce deriv-
atives ranging from oils, flours, and fertilizers to fish meal and
glues, which in some cases yield an even higher profit than the
main fisheries. Many of these derivative products also become part
of the food chain.[16] The architecture relating to their production
is not as obvious—or as magnificent—as the buildings involved in
direct food production, but it is an important element in understanding the built shapes generated by fish since it contributes to
determining the scale of the boats (when processing is done on
board) as well the physical connections between the fish landings
and their initial processing.

Fish processing consists of many operations with social, biological, and architectural implications. It transforms an animal—a fish
occupying a high rank within the food chain[17]—into a commodity.
The nature of this operation is key to this book: the industry does
not fish in order to eat what is caught but rather to take the natural
resource and give it commercial value, turning it into a product for
sale and generating economic activity. It is the commodification of
the ecosystem that propels the growth of fisheries and the construction of support facilities to host their activities—from navigation to
storage, from processing to commercialization.

12

Fishing architecture is a product of the process of value extraction that treats marine ecosystems as exploitable resources by transforming a fish into a commodity. This book begins with the mid-nineteenth century, when preindustrial fisheries were progressively supplanted by mechanized ones, changing technologies and building practices.[18] Because human demand for new sources of protein rose concurrently with these developments, fish extraction was scaled up dramatically, as was the pressure put on marine ecosystems. There were major changes to fishing and consumption practices in the decades spanning the turn of the twentieth century as food preservation techniques and facilities underwent significant transformations—the shift from artisanal salting to industrial canning of sardines is a case in point. The development of new technologies in the fisheries was, for the most part, an indirect process by which advances in other sectors (such as the engines of commercial navigation, the sonar used by the navy, the meat and dairy freezing industries, steel cables, and synthetic fibers and plastics) were progressively adopted. The architectures of fishing reacted to these changing dynamics of demand and technologies, increasing the ecological impact of fishing activities as fragile boathouses built upon dunes and in coves gave way to urban factories. Without specifying precise chronological limits, this book examines the liminal phase between the industrialization of fisheries and a still recent past. The focus on the nineteenth and twentieth centuries facilitates the study of cultures of extraction before the proliferation of industrial aquaculture and fish farming in the 1980s introduced a production model that transformed everything from the scale of global trade and fish consumption to the genetic traits of fish. Looking at the fluctuations in fish biomass in certain areas of the North Atlantic prior to this paradigm shift is useful in addressing the considerable uncertainty in our understanding of marine ecosystem dynamics and how they correlate to construction on land.

Although fisheries are often thought of as the agriculture of the sea, the biological dynamics of fish and the instability of their ecosystems don't relate to agricultural production cycles. Mining is a much more appropriate metaphor: once every fish in the sea has been extracted, there will be no fish left.[19] This tragic scenario is also a common one. There are plenty of extinct species, and if it were not for the fact that the economic profitability of a fishery often declines before the complete extinction of the target species, there would be even more. This is why Newfoundland cod—which still exist—are considered "commercially extinct."[20] There are two fundamental differences between fishing, mining, and agriculture: on the one hand, a large proportion of fisheries exploit the commons, areas without sovereignty that are universally accessible;[21] on the other, fish feed from the ecosystems they inhabit, so fishing businesses have no need to invest in products like fertilizers and fodder to secure production prior to extraction.

Common resources have always been a sensitive topic in fisheries, and there are more or less complex systems in place to manage territorial conflicts. Negotiations on maritime coastal sovereignty and the EEZ became contentious after World War II, when growing technological capacity increased pressure on marine ecosystems.[22] Fisheries regulation is, first and foremost, about access to marine resources, and other more profitable industries such as mining and international shipping also have significant interests in the sea that were discussed in the diplomatic negotiations leading up to the adoption of the United Nations Convention on the Law of the Sea (UNCLOS) in 1982.[23] Despite the United Nations convention, the legal limits of continental platforms (the extension of a relatively shallow seabed contiguous with the shoreline before the seabed falls away steeply into abysmal depths) continue to be discussed and adjusted.

"Save the products of the Land," ca. 1917–1918. Illustration by Charles Livingston Bull, United States Food Administration.

Because fish do not require feeding, its commercial extraction has an advantage over other food industry sectors, like cattle farming, where providing feed is a significant expense. An American propaganda poster printed during the Great War is explicit in advertising this strategic advantage: "Save the products of the Land. Eat more fish—they feed themselves."[24] Along with free access to the unregulated commons, this factor made fisheries a lethal activity for the targeted species. But these advantages did not create a stable business on land. While modeling economic trends relating to the exploitation of groundfish populations in the middle of the twentieth century, Scott Gordon coined the word *bioeconomy* and demonstrated the economic and environmental vulnerability of fisheries.[25] Empirical research since then has been busy proving Gordon's hypothesis that there is an initial moment in a fishery when some operators manage to extract phenomenal value from the sea with little investment; their success attracts competition in exploiting the targeted species' habitat, and, to maintain the levels of catches and revenues, operators are compelled to increase their investment in technology. As pressure upon the target species increases, the economic profits decrease in proportion to the decrease in the fish population's biomass (the reason for the increased effort), with the result that prices go up and market demand declines. Once the fish population is exhausted, exploiting it becomes a losing venture. Although it is rare for a fishery to catch the last fish of a species, it is also rare that the population levels of targeted species recover their ecological position after prolonged overfishing.[26] When a species approaches extinction, it loses its capacity to prevail in a given territory and other species take its place in the ecosystem, which makes it very hard for the declining population to bounce back.

Gordon's demonstration is crucial to understanding the architecture of fisheries. The accessibility of marine resources and high profits in the early stages of exploitation of a fishing ground create conditions that encourage significant infrastructural investment on land. Such terrestrial support helps to increase the ecological pressure upon marine populations, but soon becomes ineffective. Large-scale fishing infrastructure demands high yields from the business to secure profits to cover mortgages and maintenance, but maintaining these yields continually requires more investment in fishing and processing technologies, quickly rendering the original infrastructure out of date. Add to this equation the need for a pool of available workers—without whom most of the investments become impossible—and the means to pay them. The various cases assessed in this book show that the presence of terrestrial fishing architecture creates ideal conditions for a dangerous rise in fishing pressure. And once this pressure reaches a critical ecological threshold, the buildings—and the social impact generated by their inflexibility—act as both an excuse and a reason for continuing to expend even greater fishing effort. In most cases, the end result is the collapse of the ecosystem being exploited and the obsolescence of the buildings. In parallel, additional terrestrial investments are often made to adjust to new technologies and different fishing resources. In a pattern seen in other economic activities, when fish disappear from the sea, investors seek other fishing locations to continue the culture of extraction. The life cycle of architecture built to support fisheries—from its erection to its apogee, decay, ruin, and reconstruction—demonstrates how voracious this extractive culture is. And each step in the life of a fishing construction is matched by substantial changes in the organization of marine ecosystems.

There is no such thing as a "natural" ecosystem. The ecological amnesia that advocates the reestablishment of a perceived natural state that existed a few decades previously is called *shifting baseline syndrome.*[27] A good example of this syndrome is the legendary abundance that early European sailors found off the American coast. But this perception of a supernatural ecological wealth was merely a reflection of the poverty of European seas after centuries of overfishing.[28] If today there are less fish in the oceans than there were half a century ago, there were at that time less fish than in the preceding century, and so on.

The constant decline of fish populations is an important reference for understanding fishing architecture. Even though the decrease in fish populations became exponential during the nineteenth century with the introduction of bottom-trawling fisheries (which not only caught the fish but also destroyed their habitats by scraping all the life forms off the bottom of the sea), the activity of extracting fish from the oceans has continued to grow to astronomical dimensions. For the most part, such growth is possible because fishing grounds are physically disconnected from consumers, a split that allows tuna to be caught in southern Europe and eaten fresh in Japan, and Norwegian cod processed in China to be consumed in Brazil. This imbalanced relationship between the extraction, the processing, and the consumption of resources, as well as the geographies of revenue investment, makes of fishing architecture something that belongs more to the past than to the present. A century ago, shifting baselines of fish species were related to the transformation of coastal landscapes, and one might read in fishing architecture evidence of the interaction between terrestrial and marine ecosystems. Today, although such relations persist, they are no longer continuous or immediate.

School of skipjack tuna (*Katsuwonus pelamis*). Courtesy NOAA Central Library,
National Oceanic and Atmospheric Administration.

1 *The Cove and the Surf*

The moving of houses is Newfoundland's most spectacular architectural memory: entire buildings were rolled onto rafts and, like ships, launched upon the sea.[1] The timber houses were built on stilts above the island's rocky and uneven terrain, making them quite easy to move. A house might be relocated within a community in response to shifting familial connections or farther afield, to a different outpost, a different cove, or a distant bay. This was the case with the resettlement processes that took place between 1954 and 1975, when the provincial premier Joseph Smallwood (1900–1991) fostered several programs to concentrate hundreds of dispersed coastal settlements into "growth centers" offering social services and other public facilities. Many outport communities moved, and they brought their houses with them.[2]

An account "from the kitchen window of Mary's place" describes a not very big house of "two stories; white-washed, with a green trim round the windows and door." The resettlement funding scheme was not enough to cover the building of a new house, hence the move. Coves were crucial to mollify the ocean's harshness. "The Placentia Bay is full of islands and peninsulas. It should be easy to stay out of open water where the waves are hazardous and

Moving a house in Trinity Bay, Newfoundland, 1968. Photo: Marilyn Marsh. Courtesy
Maritime History Archive, Memorial University of Newfoundland, St. John's, PF-317.072.

unpredictable." During the journey, Mary's neighbor experienced contradictory emotions, yet she described the event with precision:

> With the pushing and hauling the house gave away within five minutes and in one loud splash it arrived on the water. ... The boat pulled taut with the load and I could see the house pull up behind us and then we were off, leaving. ... Just behind the point I could feel the tenor of the water change ... usually a bit choppier once you get out of the safety of the harbour but I felt it all the more as I turned and watched our house swaying back and forth in the waves.[3]

Following Newfoundland's confederation with Canada, resettlement policies paralleled the efforts that were made to modernize the fisheries and make the new province competitive in the context of two key postwar changes in the Atlantic, one technological and the other political. The technological change was the widespread adoption of frozen fish processing.[4] Beginning in 1954, an onslaught of international factory boats began to fish on the Grand Banks of Newfoundland's continental platform. The first to arrive was the Scottish *Fairtry*, a ship with a gross tonnage of 2,605 and a seven-ton electric trawl.[5] Soviet and West German competitors were not far behind. As historian Miriam Wright remarked, whereas "the small-boat fishery of Newfoundland was primarily a summer affair as ice-bound harbours along the northeast coast limited winter fishing," the international fleets trawled all around the island, northward toward Labrador and in all seasons.[6] Factory boats were capable of massive catches that were processed and frozen on board and quickly supplanted the inshore fisheries.[7] This led to the political change—the regulation and control of access to natural resources on continental platforms—seen in Canada's fight to extend its fishing jurisdiction to 200 nautical miles from shore, a right it won in 1977.[8]

Combined, these technical and political changes had a severe ecological impact on cod populations, leading them close to extinction.[9] In parallel, under the spell of "modernization," Newfoundland went through major social transformations. The house moves of twentieth-century resettlements ended up being the final act of an architecture that disappeared in the haste for modernization.[10] Tracking the intellectuals who directly observed the outports' shift from traditional to modern, historian Jeff Webb depicted the entanglement between a changing landscape and its underlying cultural politics.[11] Despite Newfoundland's cultural background, the transformation of cod's marine population structures was a key factor governing the reshaping of the terrestrial architecture. The light wooden outport houses and the communities around them were built for inshore cod fisheries. Without infrastructure like sewage systems, paved roads, and massive wharfs, the settlements and the houses could be packed up and moved from cove to cove in response to social changes on land and transformations in marine ecological systems.

Newfoundland's confederation with Canada in 1949 marks the chronological limit of a settling process that gained momentum in the early nineteenth century.[12] The colonization of the island began in the sixteenth century with the arrival of the first European migratory fisheries, which operated seasonally. By the seventeenth century, the English had established year-round "plantations" that shared the island with French seasonal fisheries. The settlement rate stabilized in the eighteenth century owing to several factors, among them local environmental conditions and political and economic choices made in the metropole.[13] But when nineteenth-century industrialization prompted a global population boom and an increased demand for protein-rich foods, Newfoundland saw exponential growth in population and development, with earlier plantations expanding into commercial hubs that linked dispersed outport operations.[14]

This chapter will compare the buildings of two virtually synchronous fishing cultures: the Newfoundland outports that produced salt cod and the *palheiros* settlements on the sandy Atlantic coast of Portugal that produced brined and barreled sardines. In both cases, the environment shaped the building practices: in Newfoundland, coves protected by rugged and rocky bays grant safe harbor in harsh conditions; in Portugal, the sand floor that connects land and sea allows direct access to the neighboring sardine population. Aiming for different fish species, on opposite Atlantic shores, these two forms of settlement have little in common apart from the frailty of their buildings and a significant lack of infrastructure. Both types of settlement were established with minimal means and vanished when more efficient food-processing techniques and growing commercial networks increased pressure upon fish populations. Unlike planned settlements—from Roman camps to New World colonial towns—these two forms of fishing urbanism were

Palheiros on the sand dunes, Lavos. Courtesy Arquivo Fotográfico Municipal da Figueira da Foz, box 2, 57.

24

Palheiros and beach seining boat, Furadouro, 1963. Photo: Ernesto Veiga de Oliveira.
Courtesy Arquivo do Centro de Estudos de Etnologia/Museu Nacional de Etnologia,
Direção-Geral do Património Cultural/Arquivo de Documentação Fotográfica.

shaped by logics of capital and production derived from fisheries, by their environmental and technological contexts, and, according to this book's hypothesis, by the biological characteristics of fish.

Before looking into this architecture, we must understand how and where fish live. The first key element to consider is ocean temperature, which is directly affected by water salinity and atmospheric conditions like solar heat and high- and low-pressure zones that activate the flow of energy throughout the planet. Oceanic currents add to the constant movement of warm

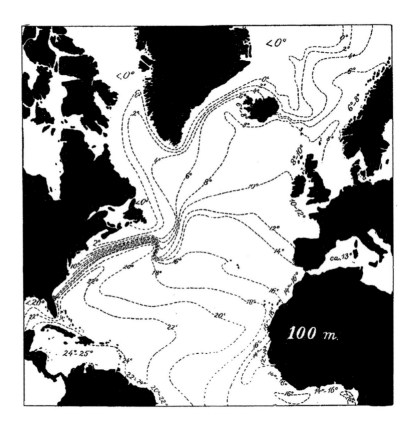

North Atlantic water temperature at 100 meters depth, in Johan Hjort, *Fluctuations in the Great Fisheries of Northern Europe*, 1914.

and cold water in tandem with the Earth's rotation. Water temperature varies throughout the water column, and deeper layers are not always the coldest. Fish metabolism is directly affected by water temperature, and a key distinction between species is their "critical temperature threshold," which defines the thermal conditions each species requires to maintain a balance between the oxygen intake its cells require to process food and its ventilation

26

and circulatory systems.[15] Not only are different fish species physiologically adapted to thrive at different temperatures, so are the phytoplankton and zooplankton that form the basis of marine food webs. For this reason, variations in ocean temperature drive the seasonal movements of fish in search of the best habitats.

The life of a fish is as dynamic as the variations in oceanic temperature, their metabolism changing as they move through the five periods of the life cycle: embryonic, larval, juvenile, adult, and senescent.[16] Each stage is characterized by physiological and morphological developments, such as acquiring fins as a juvenile, and spawning and migrating as an adult. Juveniles and adults tend to inhabit different places, and there are also major geographical divisions between populations of a single species where long-term adaptation and genetic evolution have resulted in phenotypes dependent on specific environmental contexts, which includes the ability of the fish to tolerate specific temperature ranges. For instance, Baltic and Barents Sea cod are genetically closer to North Atlantic cod than to each other, "supporting the hypothesis of a transatlantic flow of genes" and explaining dynamics within different genetic pools that vary according to interactions between fish physiology, life cycle stage, and the environmental conditions of oceanic currents and water temperatures.[17] In contrast, despite the "existence of well-defined geographic and hydrographic boundaries between the different sardine populations" of the East Atlantic region, there are connections, overlaps, and similarities between them.[18] A Moroccan and a French sardine, for example, can be distinguished by morphological characteristics rather than physiological ones, most of them attributable to the effects of water temperature and their feeding habits.

The distribution of fish species depends on the depth of the water column and the geomorphology of the ocean floor. Of the two types of fish species, pelagic fish live in the water column, often close to the surface, while demersal fish live in benthic habitats near the seafloor, an underwater environment animated by mountains, canyons, peaks, and plateaus that contribute to its ecological wealth. This topography, at depths ranging from several thousand to a few hundred meters, interacts with other environmental factors such as currents and winds to create different patterns of food availability. One of the most important hydrodynamic phenomena provoked by the configuration of the seafloor is upwelling. Interacting with forces caused by the planet's rotation, the relative position of the shoreline and the wind push the upper layers of water out to sea, to be replaced by colder waters from the deep that bring nutrients to the surface and help feed marine life. Accordingly, locations characterized by upwelling have more pelagic fish, seaweed, and vegetation.[19]

Relative to the depths of the North Atlantic, the extension of the American continental shelf is quite shallow at less than 250 meters deep. At this depth, marine flora and fauna thrive in nutrient-rich ecosystems without the assistance of upwelling. Off the coast of Newfoundland, this favorable geomorphology coincides with the meeting point of the frigid Labrador Current passing through the Davis Strait and the warm Gulf Stream moving northward. The resulting cold (but not too cold) water temperatures on the large continental shelf made the Grand Banks of Newfoundland exceptionally fertile in fish.

These environmental characteristics introduce us to two different fish: cod and sardines. Cod is a demersal fish, a carnivore that preys on other fishes, with a strong muscular mass and not much fat. Sardine is a pelagic fish that is low in the trophic chain, feeds on zooplankton, and is high in fat. Cod inhabit places where warm and cold currents merge, allowing access to the water temperature and food it requires in each stage of its life cycle. Sardines benefit from strong upwellings where the upper layers of the water column are simultaneously cold and rich with nutrients. These two fish, because of their different physiologies and living environments, are the sources of the architectural forms we are looking at.

Tender cod flesh is high in protein and low in fat, an optimal chemistry for preserving its nutritional qualities by dehydration.[20] The business of drying and salting cod started in Catholic Europe in the eleventh century and covered the entire Mediterranean Basin.[21] Fisherfolk would venture from the North Sea up the coast of Norway to Iceland, Greenland, and even Newfoundland—there are archaeological remains of Viking settlements in the north at L'Anse aux Meadows that date to between 990 and 1050—to secure a continuous supply of cod.[22] Fifteenth-century accounts often describe the sheer wealth of cod on the North American coast—a fabled abundance that drew numerous merchant-venturers far afield in the hope of tapping this rich natural resource.[23] From the sixteenth century onward, numerous British and French fishing fleets operated in the region, combining seasonal migratory fisheries (with fisherfolk based in Europe) and coastal settlements. The Indigenous people of Newfoundland were the Beothuk, who thrived within the island's forest ecosystem. Although their mixed diet included fish—which European settlers disrupted by occupying the most accessible fishing locations—the Beothuk did not develop a fishing society.[24] It was not the simple biological presence of cod in the region that prompted the coastal settlement and population of Newfoundland but the European demand for it.

The sinuous and rocky coast of Newfoundland provided a wealth of coves, a convenient setting for processing and cabotage with its patchworks of protected natural ports connected by sea. Its dry winds were especially conducive to drying fish. Cod was fished by line from large boats, and catches were either cleaned and salted on board and subsequently dried inland or landed, processed, and dried at the fish stations. Since onshore processing was more efficient and economical—it took less time and required less salt, a precious commodity—coastal settlements throughout dispersed Newfoundland coves were rapidly developed.

The small silvery capelin (*Mallotus villosus*) is crucial to our understanding of this settlement expansion.[25] Capelin is a short-lived forage pelagic species, feeding predominantly on larger copepods, and plays a key role in transferring energy through the trophic chain. The cod's diet is dominated by capelin, and fisherfolk would follow capelin shoals to find the cod as they pursue their prey. Capelin spawn both in intertidal areas of beaches and coves and in deep water, as far as the Southeast Shoal of the Grand Banks, hundreds of kilometers away from their beach-spawning relatives. The entanglement of cod and capelin determined the joint migration of both fishes from the offshore to the inshore, resulting in the opportunity for fisherfolk to catch large amounts of cod within a short period, at a time of the year with longer daylight and more sunshine for drying the salted cod. Newfoundland fishers would use seines and gill nets nearshore to catch capelin during their spring spawning migration, using it for bait, food, and fertilizer, but in many coves, however, the cod fishery's success depended on capelin's abundance.[26] In the 1990s, there was an ecological "regime shift" with a large impact on Newfoundland's capelin populations, and it has been demonstrated that after the 1992 moratorium, despite

Splitting codfish, Labrador, ca. 1950. Photo: Gustav Anderson. Courtesy Memorial University of Newfoundland, Archives and Special Collections, Coll. 429.

overleaf Petty Harbour, Newfoundland, ca. 1880–1884. Photo attributed to Simeon H. Parsons. Courtesy Memorial University Newfoundland, Archives and Special Collections, Joseph Laurence Collection, Coll. 199, 1.01.088.

29

Burin, Newfoundland,
ca. 1880–1884. Photo
attributed to Simeon
H. Parsons. Courtesy
Memorial University
Newfoundland, Archives
and Special Collections,
Joseph Laurence
Collection, Coll. 199,
1.01.036.

the absence of fishing, the predator's population kept decreasing owing to the absence of its major prey within the ecosystem.[27] Eventually, the dispersed pattern of Newfoundland coves should perhaps be attributed to the spawning habits of capelin.

The climate and interior of the island were harsh and inhospitable, meaning that prior to the twentieth century there was almost no terrestrial infrastructure.[28] Fisheries remained the prime reason for settlement, in stark contrast to the more diversified economies of the cities on the Gulf of Maine, where fishing operations coexisted with colonial trade and other productive and commercial activities.[29] In occupational terms, this meant that the formal architectural qualities of Newfoundland's habitats were predominantly dictated by the monoculture of cod fishing and processing. Connected to the outer world only by sea, they constituted autonomous communities with the sole raison d'être of harvesting a marine resource and producing a commodity that could then be traded for metropolitan goods.[30]

An 1821 atlas of French cod fisheries in Petit Nord, the northern peninsula of Newfoundland, represents the scattered migratory fishing stations at the margins of sea and land.[31] Taking advantage of the coves and the protection they offered from the harshness of the sea, the settlements consist of a few houses dotted in between large expanses of drying racks arranged in military alignment. This synthetic depiction matches late-nineteenth-century photographic records of Newfoundland outports that show predominantly wooden constructions, either anchored onto piers or built above the barren rocky landscape.[32] Apart from the conventional houses and the occasional stone building, the only terrestrial infrastructure consists of racks, piers, docks, stairways and other service constructions, all strikingly frail-looking. It was not until the twentieth century that this anti-urban, preindustrial coastal habitat would be overwhelmed by major technological, political, and biological changes.

There is no cod in the waters off the coast of Portugal where, unlike the large expanse of shallow waters off Newfoundland, the coastal profile drops abruptly twenty miles from shore. This geological configuration, in combination with the direction of the waves and the Coriolis effect caused by the Earth's rotation, produces a strong upwelling of water from the depths of the ocean to the surface. As a result, the Portuguese coast has low-temperature surface waters and a wealth of plankton, a habitat ideal for the sardine, a pelagic species. Nonetheless, this food source makes sardines particularly susceptible to environmental change. Sardines are schooling fish, moving long distances along the coast, that have a relatively short life cycle: they reach adulthood in one year, and within their first two years they can grow to more than 20 cm in length.[33]

The fatty acids of sardine flesh are well-suited to preservation in brine, a technique that developed in the late eighteenth century before being eclipsed by the more efficient canning industry during the second half of the nineteenth century.[34] The sandbars south of Porto led to the adoption of a fishing technique called *arte xávega*, or beach-seine fishing, in which a flat-bottomed boat hurdles the surf to drop a seine one or two miles offshore. The catch is later hauled in from the beach by oxen.[35] The ability to launch boats directly into the sea, without a harbor, means a fishing operation can be set up with little investment or infrastructure. As a result, various isolated sardine-fishing settlements sprouted up along the coast to provide lodging during the fishing season at beaches such as Esmoriz, Furadouro, Torreira, Costa Nova, Vagos, Mira, and Tocha. An aerial photograph of Tocha in the 1950s shows a few wooden houses aligned across the dunes, with a long umbilical cord connecting the coastal settlement to the main road and railway systems a few kilometers east. It wasn't until seaside tourism brought in a different temporary residential clientele that a corresponding infrastructure was developed.[36]

With their rickety wooden *palheiros* erected on stilts above the sand, such settlements were not so different from their cod-fishing

counterparts in faraway Newfoundland in their direct relation to the marine resources, their reliance on communication networks, and the absence of major infrastructure on land. But unlike Newfoundland, where the coves and bays are safely navigable by larger commercial boats, the roughness of Portugal's shoreline and sea make having port facilities or a direct connection to land-based infrastructure necessary to transport catches to the main processing facilities. And unlike cod, which can be preprocessed on board and wait till further processing due to its low-fat meat, quick transportation and processing is a necessity for fatty sardines, which otherwise will decay rapidly.[37]

The examples of Newfoundland cod-fishing and Portuguese sardine-fishing settlements highlight how architecture occupies the intersection between an ecological system and its economic exploitation. For a fishing business to succeed, the fish must be properly handled and strong forward and backward linkages put in place to transform the resource into profit.[38] Forward linkages are industries that transform the fish into commodities (by processing and commercializing it); backward linkages are the infrastructure required to operate such industries (such as facilities that

34

produce hardware and other necessary equipment, power plants, and railroads for transportation). Their cost determines the rent extracted from natural resources, and effective linkages can generate the most wealth. The linkages relied upon by fishing settlements in Portugal and Newfoundland were fragile.

A more successful counterexample can be found in Iceland, which from the nineteenth century onward grew a more diversified economy.[39] Although the economy was initially based on fishing, government and currency policies supported the development of complementary activities that developed into autonomous sectors such as banking, shipbuilding, and business. That is very different from the situation in nineteenth-century Newfoundland, where the majority of the population were kept dependent on fish catches by means of the practices of direct loans—trading fish for credit—and diversification within the same sector—introducing herring and seal fisheries as alternatives to cod.[40] Portugal's coastal settlements did not develop diversified economies either, but for a different reason: they became seasonal seaside resorts. It wasn't until later that an important harbor with industrial infrastructure was built up the coast to the north, resulting in what might be referred to as sardine urbanization.

The absence of terrestrial infrastructure—which makes the shoreline settlements of Newfoundland and Portugal surprisingly similar—is due to politics. The fact that most of Newfoundland's shipbuilders, merchants, and bankers were operating from Europe resulted in perennial uncertainty concerning the administration of the territory. It wasn't until the 1950s, after the former colony was integrated into Canada, that there were significant changes in settlement patterns. In Portugal, strong investment in the canning industry made it possible to build port facilities and large factories, resulting in fishing settlements more urban in character than those built to serve the beach-seine fisheries that produced brined sardines. As in Newfoundland, soft political intervention and weak linkages granted these older fisheries an elementary fishing architecture, whose construction depended less on technological investment than on an intensive exploitation of natural resources. With low capital investment, the ecological conditions of fish habitats and biological characteristics of fish species—for instance, water temperature and critical temperature threshold—were key in determining the architectural qualities of the settlements and their modus operandi.

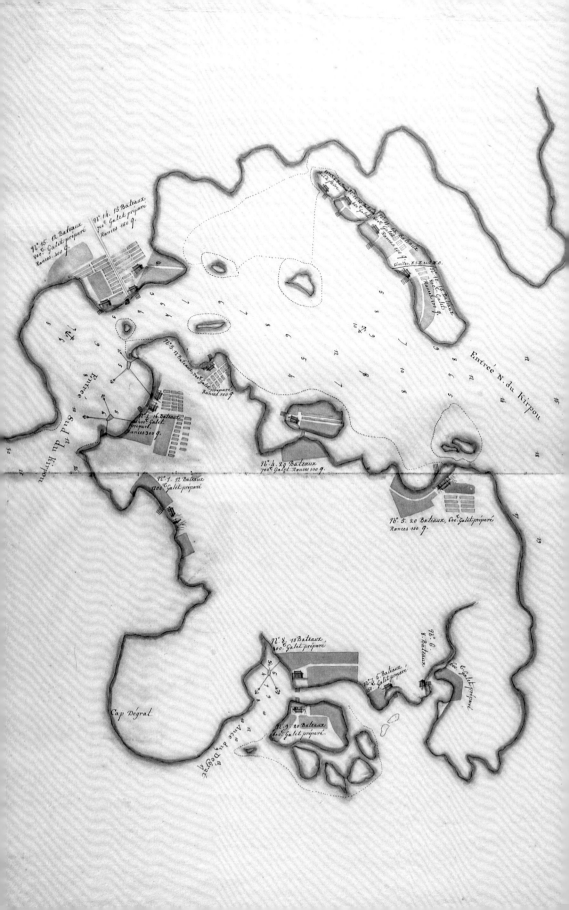

Entrée N. du Kirpou

Entrée du Sud du Kirpon

Cap Dégrat

Nᵒ 15. 12 Bateaux
800ᵗ Galet-préparé
Rances 300 q.

Nᵒ 14. 15 Bateaux
700ᵗ Galet-préparé
Rances 150 q.

Nᵒ 13. 15 Bateaux
700ᵗ Galet. 150 q.

Nᵒ 3. 11 Bateaux 700ᵗ Galet-préparé
Rances 100 q.

Nᵒ 2. 16 Bateaux
1000ᵗ Galet-
préparé
Rances 300 q.

Nᵒ 1. 12 Bateaux
800ᵗ Galet-préparé

Nᵒ 4. 29 Bateaux
700ᵗ Galet. Rances 100 q.

Nᵒ 5. 20 Bateaux, 600ᵗ Galet-préparé
Rances 100 q.

Nᵒ 8. 10 Bateaux,
600ᵗ Galet-préparé

Nᵒ 6.
8 Bateaux
600ᵗ Galet-préparé

Nᵒ 7. 5 Bateaux
150ᵗ Galet-préparé

Nᵒ 9. 20 Bateaux
600ᵗ Galet-préparé.

Anse du Dégrat

Although the architectural forms of Newfoundland fishing operations changed according to different coves, bays, and epochs, whether responding to seasonal or permanent operations, they nonetheless relied on four fundamental wooden structures that punctuated what are commonly named "rooms." On the wharf, there was the "stage," an amphibious platform where the fish were offloaded and processed; adjacent to the stage was the "store"—and sometimes various "stores"—also used for salt and fishing gear, repair tools, etc. Sometimes stores might have served as temporary sleeping quarters, but generally sleeping was done in an adjacent house, the "cook room," and the stage would expand into "flakes," very large surfaces laid over stilts where the fish was dried.[41] These key elements of the sea-land interface were constants, from the first migratory stations of the fifteenth century to the communities of the early twentieth century. Although there were differences and variations between early English settlements and French migratory stations, between fishing outposts and settled colonies, between sixteenth-century and nineteenth-century cultures, these are mainly evident in the complementary elements of housing and storage rather than in the fishing room itself.[42] Most business transactions took place at the premises of the merchant or fish buyer, and not every cove had a merchant. Where there was one, the buildings and architecture would be noticeably different. The stages would be larger, and complementary facilities such as salt stores, cooperages, carpenter's workshops, and retail shops could be found. Flakes would overtake the cove landscape. Thus, taking into account all its multiple forms and variations, the Newfoundland fishing rooms might be considered to be cod architecture.

In Tilting, on Fogo Island, there was no fish merchant; the nearest was located in Fogo Harbour. Yet the architecture of this cove has been studied in detail by Robert Mellin, and it is quite characteristic of Newfoundland's fishing rooms.[43] The settlement wraps around a cove sheltered from the rough ocean by the small Pigeon Island. Inside the bay, a rocky peninsula separates the inner pound from the harbor. The natural form of the protected cove is domesticated by construction to become the town's most defining architectural artifact. Around it are fishing stages that interface between land and water, each placed for the best possible access to fishing grounds and belonging to a different fishing family. The houses are located

facing Cod-fishing migratory settlements, Quirpon Bay, Newfoundland, in Claire-Desire Letourneur, *Atlas*, ca. 1821. Courtesy Memorial University of Newfoundland, Archives and Special Collections, Coll. 477.

in and among the stages but in more protected positions. As in other permanent settlements, houses constitute the built core of the community, complemented by a handful of service buildings: a lighthouse, a fire hall, a post office, a church, and a parish house. The main façades of the houses, or at least the most decorated or painted ones, face the bay, confirming its function as the community's central space. Until not long ago, houses had infield gardens for growing vegetables, and clusters of outfield gardens were a short walk away in an area punctuated by stables, hay-houses, milksheds, and buried root cellars. This outer zone, with a cemetery at its northern edge, effectively extended the community's imprint upon the landscape. An extensive network of fences was meant to limit the circulation of the abundant animal population, define its main paths, and unite its disparate elements into a coherent whole.

Fogo Island's rugged topography is marked by a complex geometry of bare eroded rocks. Most photographs of Tilting render the water as a smooth if not graceful surface, its calm accentuated by the bay's protected character. But Newfoundland's ocean is a rough place, with winter storms battering the intertidal areas and sometime icing entire coves. In such a demanding environment, Newfoundlanders developed a light and reversible approach to wooden building that avoided the need for masons, whose tools and skills were rare, and the heavy work required to shape stone wharfs.[44] Nonetheless, transporting wood to Tilting from its source on Fogo's southern shore was not easy. This was usually done in winter by horse-drawn sled, when fisherfolk were on standby and smooth shortcuts over frozen ponds eased the task.

The timber was used to construct wooden houses, stages, and flakes on stilts, forming sinuous platforms that seemed to float over water and stone. Often, especially on stages that project far into the bay, the stilts were secured by foundation cribs, the crossed timber logs of which formed "ballast lockers" weighed down with rocks.[45] Strouters and head bedding at the water end protected the structure and allowed boats to be moored. Tilting's stilts and cribs raised a vast platform upon which life and activities took place, an elevated world defined by a circulation network of bridges connecting stages, flakes, and rocks, occasionally overlapping at points where the intricate paths of access to properties warranted it. The extensive surfaces allowed ample space for fish to dry on bedding made up of thin wooden rods set slightly apart, with thin paths of planks laid overtop for walking and for separating different stocks of drying fish. The gaps between the rods allowed for good drainage and air circulation, a requirement due to the continuous washing involved in fish handling, processing, and drying. They also filtered light into the spaces below, creating spectacular underworld universes.[46] This configuration, visible in photographs and still evident in traces remaining today, is not much different from that shown in depictions of the outports dating from the seventeenth to nineteenth centuries, suggesting that constructive practices in fishing stations persisted even with as new building cultures were introduced.[47]

The end of the Napoleonic Wars in 1815 marked a turning point in Newfoundland's history. The local dispute between British and French fisheries ended, and the population and the economy became progressively more stable.[48] As growing European demand for cured cod fueled fishing activity and encouraged immigration, residents replaced the former migratory fisherfolk. The population of Newfoundland increased tenfold from the late eighteenth century to 1884, when 75,000 people inhabited the island, only 15,000 of whom were in St. John's. The vast majority were dispersed among countless outports and small communities along

Below the fish flakes, Newfoundland, ca. 1950. Photo: Gustave Anderson. Courtesy Memorial University Newfoundland, Archives and Special Collections, Coll. 429.

the shore.[49] Their distribution pattern reflects the location of rocky coves offering safe harbor for fishing boats and the presence of nearby inshore fishing grounds, where even small boats, requiring less capital investment, could bring in good catches.

The first ecological consequence of this settlement pattern was the depletion of specific cod populations and a resulting loss of genetic diversity within North Atlantic cod long before its collapse. Already in the nineteenth century, Newfoundland fisherfolk were aware of separate inshore and offshore cod populations, the inshore subdivided into bay and headland groups, the offshore into migrants and residents. These similar cod had different habits: bay populations "spawn and overwinter in the deep arms of bays," in contrast to those that "overwinter in deep water off headlands" and capes, while offshore migrants "overwinter on the edge of the continental shelf and migrate inshore to feed in late spring and early summer," unlike offshore residents, which "do not migrate inshore."[50] Catches contained a mix of inshore and offshore populations, and their relative wealth concealed the overfishing of different inshore populations. A rule of thumb for sustaining cod populations is to limit catches to 18 percent of fish biomass, thereby avoiding a mortality rate higher than the population's capacity to reproduce.[51] Since the inshore populations were being fished indiscriminately from offshore ones, and landings were increasing, the ongoing massacre went unnoticed. Thus the sprawl of settlements along the Newfoundland coast, and the consistent fishing effort they represented all along the shoreline, is at the root of the systematic depletion of the genetic lineage of inshore cod populations.[52]

With inshore fish populations depleted by the second half of the nineteenth century, Newfoundland's fisherfolk had to increase their efforts and venture farther from shore to maintain their average fish landings and income, a process studied by Sean Cadigan and Jeffery Hutchings.[53] One way to do this was by using more efficient fishing gear. The old hook-and-line method was gradually replaced by traps, trawl lines, seines, and gill nets,[54] increasing the size of catches. Another way was to diversify catches, with many former cod fishers pursuing herring and seal. Seal hunting in particular required schooners and larger vessels that could travel long distances, and this in turn brought fishers to unexploited offshore cod fishing grounds. As the years went by, Newfoundlanders were fishing farther and farther north up the coast of Labrador, between the Strait of Belle Isle and Sandwich Bay and even as far as Hopedale. But the fish populations in these areas eventually declined as well, and although some locations doubled their harvest capacity between 1857 and 1891, their catches dropped more than 30 percent.[55]

One sign of overfishing is the disappearance of large fish and an overall decrease in average fish size. The need to travel farther to catch smaller fish, and less of them, had dire consequences for Newfoundlanders. Fish harvested far from processing facilities are not as fresh when the curing process starts, and since the best cured cod is made from large, thick-fleshed fish, smaller fish result in a lesser final product. As a result, as the fishing grounds expanded, the quality of Newfoundland's cured cod declined, and so did its market price. Producers countered the devaluation of the product by increasing the volume of exports, which only compounded the problem by putting additional pressure on dwindling fish populations.

The field of knowledge that seeks to understand the interplay of population, economy, and natural resources is called bioeconomics, or bionomy, and many of the tools it uses to analyze the dynamics of these relationships are based on concepts drawn from fisheries management. Bioeconomic data relating to fishing activity is expressed mathematically and correlated to other biological indicators, such as population, and to economic performance. In a 1954 paper that remains a classic of the field, Scott Gordon asserts that the "natural resources of the sea yield no economic rent" and demonstrates how their use eventually leads to a profit margin of zero.[56] Focusing on the economics of demersal fish, he draws on biological evidence—such as the characteristics of fishing grounds and how location is an indicator of certain morphological traits within a given species—to conclude that after a limited period of economic exploitation, the productivity of a fishing ground will eventually drop to the point that there are no longer enough fish for any profit to be made. This bioeconomic effect is accelerated by the common nature of the resource wherever competing fleets fish the same locations, as occurred in the waters off the coast of Newfoundland.[57]

Load of fish, Newfoundland, ca. 1940. Photo: Holloway Studio. Courtesy The Rooms, A 42–44.

By the late nineteenth century, Newfoundland's inshore fishery was struggling to keep up with the profits and the technological advances being made by Atlantic competitors fishing offshore. Steadily increasing pressure on the inshore cod populations meant the average age of catches kept dropping, and along with it the quality of the dried salted product and, ultimately, its price. In early-twentieth-century Newfoundland, salt cod was of lower quality than its European counterpart produced from fish caught by migratory fisheries in the offshore Grand Banks. Even more devastating to Newfoundland's outports were the technological innovations that occurred in the fishing sector. To keep up with the competition, Newfoundlanders would have to adjust from inshore to offshore fishing. This required costly upgrades to fishing equipment and processing facilities, which didn't come until after Newfoundland joined the confederation after World War ii, and subsidies were made available in support of the industry. Moreover, the shift from salt cod to frozen products changed the scale and the logistical networks of commercialization, a step that meant leaving behind the exhausted inshore cod populations to go after the still-plentiful offshore populations.

The cornerstone of Newfoundland's outport settlement structure was the economic exploitation of cod. And although the shift from inshore to offshore fishing can be justified economically, the fact is that it was based on ecological conditions. Since fish often reproduce at high rates, a fishery can enjoy durable profits as long as the harvested catches do not surpass the population surplus. But by the late 1940s, Newfoundland's inshore cod populations could no longer cope with the crescendo of pressure placed upon them as the growing outports expanded their activities regionally beyond all reasonable limits. Refocusing on offshore fisheries required new technologies and methods for extraction, processing, and distribution. Economic and architectural adjustments were

necessary, prompting not only new architectural forms but also the resettlement policy of the 1960s and its house moves, a radical architectonic operation. And despite all the effort expended, it was not long before the offshore cod populations succumbed as well. By the late twentieth century, all of Newfoundland's cod populations were on the verge of extinction.

Abundant sardine populations have long lived off the coast of Portugal,[58] but there was no relevant sardine industry there until the second half of the eighteenth century, when Spanish entrepreneurs,[59] eager to expand their profitable business fishing the Galician fjords, brought the *arte xávega* technique to the sandy stretches south of Porto. Over the next two centuries, coastal settlements formed by wooden *palheiros* sprouted up along the sand dunes, until competition for the same fish population from Leixões and Póvoa de Varzim rendered the beach-based industry obsolete. The *xávega* boats could not compete with the harvesting capacity of seiners launched from these northern ports, whose strong harbor infrastructure permitted larger boats and supported canneries operating at capacities that supplied an expanding export market. Although it seems obvious that such economic and technical factors encouraged the appearance and disappearance of the *palheiros* settlements,

Palheiros on the sand dune, Furadouro, ca. 1950. Photo: Mário Almeida. Courtesy Centro Português de Fotografia, PT/CPF/EA/ rf-30-3-3.

it is worth investigating how intimately their existence and form were connected to the surrounding environment, in particular the marine ecosystem whose conditions allowed the sardine to thrive.

The relatively young shoreline south of Porto—formed over the Vouga estuary during the last 500 years—signals the ecological continuum between sea and land and connects the sardine's habitat to the *palheiros* of their human predators. The coastal environment is sensitive and dynamic. Strong north winds formed by the temperature differential between sea and shore blow year round. Sand is pushed by the wind and dragged by the water, with waves pulling the dunes back and forth in seasonal movements and the varying configurations of the sand shifting the surf break. The sandy, exposed seafloor is inhospitable to most marine flora and fauna, except for pelagic species like sardines. The dunes were not much more welcoming to human settlement. Extensive landscaping was done to counter the effect of the wind that kept the sand in constant flux, including planting massive stands of pine trees. The establishment of fishing settlements along the coast was thus anything but natural. Establishing a population on the dunes facing the surf break was economically motivated, part of a plan to exploit a natural resource by transforming it into a commodity to be traded in distant markets.[60]

There are several environmental rationales for the elevated form and wooden construction of the *palheiros*. The instability of the sand is one of them. Raising the houses on stilts was a practical solution to avoid the threat of blowing sand accumulating over the north façades. Also, with the shoreline in permanent movement, elevated houses were easier to relocate if high tides came too close. Another is the accessibility of building materials. The timber used in construction was presumably sourced from the nearby pines planted to stabilize the sandbars from the eighteenth century on. Brick and stone were seldom used before road infrastructure was developed to bring them in (along with seasonal tourists and the higher standards of comfort they expected).

The rough surf on the west coast of Portugal comes from exposure to energy dissipated by the Azores High in the form of wind power that is transferred into waves that crash upon the shore. The sandbars south of Porto have neither rocky coves where boats can be moored safely nor river mouths deep enough to accommodate large hulls, which makes getting over the surf break the most important movement in accessing the shore fisheries. The key to

success is a boat with a flat bottom that can slide from sand to water, like a twentieth-century surfboard, and a prominent bowed prow that can pierce the wave and divert its power to either side.[61] These are the key features of the relatively small, light "half-moon" boats, the fundamental sea-land interface that allowed the beach-seine fisheries to thrive. Once over the surf break, the crew would row a mile or two out to deploy a seine—the *xávega*—connected to a rope anchored on the beach to be later dragged in by oxen.[62] The technique requires a sandy seabed, since a rocky bottom would snag and damage the expensive nets.

Thus the half-moon boat that carried the *xávega* conquered the surf to establish conditions favorable to coastal settlement. This began in the eighteenth century with temporary camps for a seasonal population whose movements echoed those of the sand-bars. Permanent settlement quickly followed in areas like Espinho and Furadouro, quite close to the established town of Ovar. Others took longer: Mira did not have a regular population until the 1860s, when the fishing community was able to work at nearby farms in the off-season.[63]

Although the *palheiros* are best known as fisherfolk's houses, they also served as stables for oxen and their fodder as well as

Oxen pulling the beach seine in *arte xávega*. Courtesy Biblioteca Central de Marinha Arquivo Histórico, FG/009-11-002/013.

45

warehouses for fishing gear, salt, and barrels of sardines in brine. Such all-wood construction is not common in Portugal's vernacular architecture, where structural wood usually appears in combination with stone or brick.[64] The wooden *palheiros* are a specific architectural expression of the *arte xávega* fisheries and their unique strategy to confront the surf. A thorough ethnographic survey of the structures still standing in the 1960s conducted by Ernesto Veiga de Oliveira (1910–1990) and Fernando Galhano (1904–1995) shows that the construction techniques used to build them were quite varied.[65] The most sophisticated of them had the wooden stilts supporting a structural floor frame, others were built upon crossed beams, and many had the stilts continue as support columns for the roof (making it impossible to move the house if ever needed). They were painted with the same red lead paint used to waterproof the half-moon boats, and this deep color made them stand out against the white dunes.

It is tempting to connect these wooden houses to the wooden boats produced in nearby shipyards. However, while the specialized boatbuilders used top-quality materials, the *palheiros*, which were often closer to shacks than proper houses, were built of lower-quality timber by less qualified workers. How much did a boat cost? And a *palheiro*? Such comparison should not be taken at face value, since a seaworthy boat also needed fishing gear, a crew, animals to haul the nets, and the fodder to maintain them. Each boat would be associated with many houses, and its size reflected the social organization in which it played a key role. In any case, the answer to the question of cost, which I do not know, might give a rough insight into the relative importance of the settlement and the fishing gear.

Fire and water were constant threats to the *palheiros* settlements. The sand dunes they were built on were anything but stable, and coastal erosion combined with strong tides occasionally washed buildings away. Despite the humid seaside air, the labyrinthine agglomerations of *palheiros* were prone to frequent and devasting fires. Furadouro, the largest of these seaside communities, burned down twice, in 1881 and 1892, prompting planning authorities to impose an orthogonal grid in 1904 to regulate sprawl and make it easier to control future conflagrations.[66]

The rustic and informal character of these settlements relied on their remoteness. But access to transportation was important to support the growth of fishing activities. At Mira and Furadouro,

Splitting, washing, and cooking sardines at Feu Hermanos canning factory, Portimão, Algarve, ca. 1960. Photos: Júlio Bernardo. Courtesy Museu de Portimão/Centro de Documentação e Arquivo Histórico, MP-D2-23B, MP-D2-6B-B, MP-D2-9B-A1.

Hauling sardine seine, Portimão, Algarve, ca. 1960. Photo: José Anacleto dos Santos Dias. Courtesy Museu de Portimão/ Centro de Documentação e Arquivo Histórico, JASD-D1-74.

relatively short roads were cut through the pines to meet the main trading routes. Torreira and Costa Nova had direct access to Vouga's lagoon on the inland side of the sandbar, providing an efficient water exit for the salt-barreled sardines. A canal connecting the lagoon to Porto's harbor was planned in the eighteenth century,[67] but the two remained unlinked until the railway was inaugurated in 1864. Aiming to boost the *arte xávega* fisheries, this new infrastructure brought tourists instead. In Furadouro, guesthouses opened, and summer rentals of *palheiros* began shortly after the macadam road was built around 1869. The road also brought stone and brick from which houses were built along the main street, developing commerce and welcoming a population not too keen on the scent of sardines. The result was a strategic separation within the community: fishing activities were pushed to the south end of the beach so the direction of the wind would spare the tourists at the north end from the smell. The same happened in other settlements, and the shack-like *palheiros* were slowly transformed into charming summer cottages.

The second half of the nineteenth century brought an even greater challenge to these settlements in the form of competition from canned sardines. Portugal's first canneries were established in Setúbal, south of Lisbon, by the mid-1850s. The canned-sardine

overleaf Adão Polónia & Cia. canning factory, Matosinhos, ca. 1943. Photo: Alvão. Courtesy Centro Português de Fotografia, PT/CPF/ALV/04891.

industry got a boost from the opening of the railroad network in 1864, but the most important factor in its growth occurred in 1880, when a sardine "regime shift" off the Breton coast shuttered what had been a very successful French industry.[68] When the fish simply vanished, the French industrialists recognized Portugal, with its emerging canning business, as a promising outpost in which to invest their capital, technical know-how, and access to commercial networks. From the initial base in Setúbal, by 1889 a major canning factory was operating in Espinho, and in 1901 another had opened in Ovar by the train station, a prime location from which to exploit the Furadouro fisheries.[69] These factories produced a canned product at a larger scale and a lower cost than brined sardines. Consumers also preferred it: not only were the canned sardines tastier and more nutritious, they were also less expensive. Brine and barrel sardines lost market share and value, and both *palheiros* and the *arte xávega* started to lose their raison d'être.

The rising popularity of canned fish led to market demand far beyond the capacity of the *arte xávega* fisheries. Instead, the canneries were mainly supplied by seiners operating from the rocky northern harbors of Leixões and harvesting the same sardine population being fished from the beaches. While the purse-seine method they used requires a greater capital investment in boats, nets, and crews, it also generates much larger catches. And these large catches were landed in close proximity to the sprawling industrial complex of Matosinhos, a facility developed by the harbor itself to process and distribute the commodities internationally. It was not long before the canning factories of Matosinhos took the lion's share of sardine landings. Even so, there were subtle reasons for preserving the older techniques as a means to absorb occasional fluctuations in supply. Since canning sardines requires the fish to be fresh, and canneries have a daily maximum output, surplus fish were brined instead. When there were fish shortages, canneries would buy fresh fish from the smaller *arte xávega* operations.[70] Nonetheless, this apparent complementarity did not prevent the seemingly inevitable obsolescence of the *arte xávega* fisheries.

Arguably, the underserved *palheiros* settlements on the sand dunes owed their existence to sardine physiology and its suitability for brining. Without sardines and their pelagic behavior, there would be no *palheiros*. But it was the technologies of brine and barrel processing and especially of the flat-bottomed boats for *xávega*, not the fish themselves, that launched a fishing culture and gave rise to the appearance and spread of *palheiros* architecture. And once that fishing culture industrialized by shifting to seiners and canning, so did the pressure on the marine ecosystem. There were already alarming signs of overfishing by seiners in the 1930s, in contrast to the negligible ecological pressure the *xávegas* had put on the sardine population. Nonetheless, the *palheiros* disappeared not on account of environmental or ecological causes but because of the technological shift that occurred when canned sardines supplanted brined and barreled ones. And along with the end of the *arte xávega* economy and the *palheiros* architecture associated with it went the sense of environmental continuity they created by inhabiting the inhospitable sand dunes and exploiting the marine ecosystem just over the surf break.[71]

2   *The Whale and the Shore*

On April 20, 1861, the week after the outbreak of the American Civil War, *Vanity Fair* published a cartoon showing a "Grand Ball Given by the Whales in Honor of the Discovery of the Oil Wells in Pennsylvania." Despite the war and its anxieties, we see sperm whales in fancy dress celebrating, dancing, and drinking champagne, imagining a more relaxed future. They celebrate because mineral oil was an effective low-cost alternative to whale oil, a difficult-to-extract product that was rendered from the mammal's blubber. But in the end mineral oil did not even have to compete with whale oil. Whales themselves had become scarce, compromising Atlantic whaling operations and the wealth of cities such as Nantucket and New Bedford. Although that same year of 1861 saw the carnage of the "Golden Age" of American whaling approaching its end, other countries and economies continued to pursue whales using technologies that were significantly more efficient than the rudimentary preindustrial fishing techniques of the Nantucket fisherfolk, allowing them to catch in a single year what New England whalers could take in a decade.[1] Still, nineteenth-century American whaling provides a compelling illustration of an imbalance between a targeted population of sea inhabitants and its built counterpart.

GRAND BALL GIVEN BY THE WHALES IN HONOR OF THE DISCOVERY OF THE OIL WELLS IN PENNSYLVANIA.

"Grand Ball Given by the Whales in Honor of the Discovery of the Oil Wells in Pennsylvania," in *Vanity Fair*, April 20, 1861.

The encounter between architecture and whales usually happens in the realm of metaphors, most leading back to Jonah repenting in the leviathan's womb after having been thrown overboard.[2] However, it is not the structural performance of the colossal span of the whale's rib cage that attracts the architectural mind. Instead, whales exist on an oceanic scale, challenging the view we have of them from the shore. While their mass is incomparable to the human body, they move on a planetary scale that was barely known before industrialization. Whales are aquatic mammals, but they share the ocean environment with fish and were (and still are) hunted by humans in ways equivalent to fisheries. There are many species of whales with distinct biological features, differing in their feeding habits, migratory routes, and habitats. The same broad category combines cetacean animals as different as baleen whales (*mysticetes*) with their filter-feeding system and whales with teeth (*odontoceti*) that allow them to feed on fish or squid. Not long ago, whales were found all over the planet and traveled thousands of miles throughout the oceans, a mobility that highlights the relative immobility of whale fishing architecture. To chart a taxonomy of world whaling, Randall Reeves and Tim Smith catalogued over a hundred different operations dating from antiquity up to today, both for subsistence and commercial purposes, into eleven eras that reflect changing patterns of whaling geography, modes, and methods. They link each operation—whether shore-based, coastal, or long-distance—to a specific ethnic group and a precise terrestrial position, a location that would affect multiple interrelated whale populations.[3] The history of American whaling, an industry active from the mid-seventeenth century until the Great War, is telling of several aspects of the relationship between whales as natural resource and the architecture of their exploitation. The goal of this epic hunt was whale oil, not food.[4] Especially between 1815 and 1861, when the price of whale oil was at its peak, the massive scale of the slaughter at sea was paralleled by an urban boom in the towns that launched the whalers.

Moby Dick, the star of Herman Melville's (1819–1891) novel, was a sperm whale or cachalot (*Physeter macrocephalus*), whose oil—the spermaceti—was a highly prized nineteenth-century commodity.[5] The book's literary success contributed to the epic status of American whaling, which used preindustrial fishing practices to serve industrialized urban markets. American whale hunts were harsh long-distance sailing ventures, lasting two to three years and circumnavigating the globe. The gigantic creatures were chased and harpooned by whalers from twelve-man shallops, small boats that were pulled along at fierce speeds by the wounded whales to which they were attached. Once killed, a whale would be towed to the main boat, tied to the outboard side, and processed in a daylong operation that stripped it of blubber and other valuable parts. What remained of the carcass was released into the ocean. The blubber was processed into oil and barreled, and the hunt resumed until the crew had filled enough barrels to return home. Factories back in the whaling ports would then refine the processed whale products into commodities such as candles, oil, and corset boning.

For the most part, the Americans targeted five species: right whales (including the southern *Eubalaena australis*, the North Atlantic *Eubalaena glacialis*, and the North Pacific *Eubalaena japonica*), the sperm whale, the bowhead whale (*Balaena mysticetus*), the humpback whale (*Megaptera novaeangliae*), and the gray whale (*Eschrichtius robustus*).[6] And often, when their preferred prey was scarce, they also hunted medium-size toothed whales, such as the killer and pilot whales (*Orcinus orca* and *Globicephala*), belugas and narwhals

(*Delphinapterus leucas* and *Monodon monoceros*), and even porpoises (*Phocoena*). Many of these gigantic animals are now on the verge of extinction: as an example, it is estimated that there are only about 300 North Atlantic right whales remaining.[7] American whaling activities also had a detrimental impact on the populations of other species, notably the turtles the whalemen harvested for fresh meat during the long voyages.[8]

The history of Nantucket, Massachusetts, a small sandy island in the bay south of Cape Cod and a major center of the whale oil industry, provides valuable evidence of some important aspects of fishing architecture.[9] Whaling first thrived on the island in the late seventeenth century. Following a slump that began during the American Revolution, when the business was devastated by disrupted access to resources and markets for trade, it bounced back in 1815 when the end of the Anglo-American war brought peace to the seas and a corresponding increase in the exploitation of marine resources. Oil was in high demand to lubricate machinery during the Industrial Revolution, and Nantucket had the experience and knowledge to lead the industry, putting it ahead of the whaling ports along the Eastern Seaboard. But by the mid-nineteenth century, several architectural factors had given the neighboring continental port of New Bedford a leg up. This began Nantucket's slow decline, which was nudged along by the American Civil War (1861–1865) and the synchronous usage of rigs for extracting mineral oil. Soon afterward, crude oil and fossil fuels replaced whale oil as a source of energy, and the business faded away.

It is hard to untangle the reasons why a thriving business flourished on such an inhospitable island. First of all, whaling has a long history on Nantucket. The ecology of the bay it lies in is hospitable to whales, and the occasional presence of beached whales on the island probably incited Indigenous populations to take advantage of the carcasses. Second, the island is well situated to support successful inshore and offshore fisheries. But whaling became a long-distance activity, especially after British settlers took over from the island's strong, stable Indigenous society. The British settled in the town of Nantucket, then named Sherburne, on the northern shore, facing the continent with sea access west of a long sandbar that formed an inner bay. The Indigenous inhabitants occupied the island's east shore, in community groupings of which Siasconset would become the most stable and long-lasting.[10] This leads to a third factor: despite tensions, Natives worked with settlers in

Five species of whales sought by American fisheries. From top: gray, humpback, sperm, bowhead, and right whales, followed by a whaler.

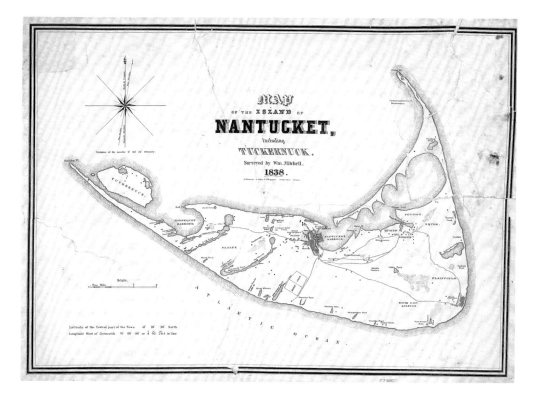

the industry (to the benefit of the latter), and the combination of Indigenous and settler knowledge of whaling operations might be one reason why the island took command of the American whaling industry.[11] Another might be that on an island where local resources were scarce and access to trade was hampered by difficult navigation, whaling proved a profitable and viable métier. The degree of specialization required also gave Nantucket a competitive advantage because of its focus on whaling. Although the industry was hit hard and slow to recover from economic crises, continental rivals, mostly located in New Haven and Long Island, were also distracted by other economic activities.

Nantucket's prominence in whaling was likely fostered by the specific limitations of its environment. A group of English settlers arrived in 1659 and negotiated an advantageous land arrangement with the Indigenous population that would support farming. But English mismanagement quickly decimated the necessary resources. Nathaniel Philbrick notes that seventeenth-century records are "filled with measures to conserve and protect grass and farmlands as well as trees," meaning that in a very short period

The Island of Nantucket. Map by William Mitchell and Ephraim Bouvé, 1838. Courtesy Harvard University, Harvard Map Collection.

61

EARLY WHALING STATION ON NANTUCKET (Conjectural)

Showing a Six-Man Boat's Crew Ashore

Section

Fish Racks

Whale Hut

Drawn & Copyright'd by H.C. Forman

Conjectural reconstruction of early whaling station. Drawing by Henry Chandlee Forman, 1961. Courtesy University of Maryland Libraries, Special Collections and University Archives.

the island's forests had been depleted by lumbering, and the land had been exhausted by intensive agriculture and the overgrazing of sheep.[12] This prompted the island's population to look for other profitable activities. So these ecological limits, combined with the fact that whale oil was not taxed by the Province of Massachusetts Bay (to which Nantucket was subject after 1692), must have contributed to the island's somewhat unexpected new line in whaling.

In his *Letters from an American Farmer*, J. Hector St. John de Crèvecœur described the shore whaling stations he found when visiting Nantucket in the 1770s, and the distinctive lookout masts that marked the island's sandy flats:

> The south sides of the island from east to west, were divided into four equal parts, and each part was assigned to a company of six, which though thus separated, still carried on their business in common. In the middle of this distance, they erected a mast, provided with a sufficient number of rounds, and near it they built a temporary hut, where five of the associates lived, whilst the sixth from his high station carefully looked toward the sea, in order to observe the

spouting of the whales. As soon as any were discovered, the sentinel descended, the whale-boat was launched, and the company went forth in quest of their game.[13]

The masts allowed whalers to catch sight of telltale spouts and whales on the horizon and effectively expanded Nantucket's territory several nautical miles into the sea.[14] Along with the hut for the six-man crew and some fish racks, they were a key component of the early Nantucket whaling station as shown in a reconstruction by architectural historian Henry Chandlee Forman based on Crèvecœur's description and a rough sketch in the island's *Proprietor's Book* of 1776.[15] These first settlements were the seeds of Siasconset, an agglomeration that combined coastal cod fisheries with shore whaling. According to the island's early historian Obed Macy, the industry peaked in 1726, with eighty-six whales captured.[16] By the mid-1750s, shore fisheries were declining, and more crews were sailing in sloops on monthlong voyages.[17]

Offshore whaling made Nantucket's lookout masts obsolete. The island's new vertical landmarks were lighthouses, such as the ones at Brant Point (1746) and Great Point (1784), both of which have been rebuilt several times.[18] Their beacons facilitated navigation and were thus important to long-distance fisheries, supporting Nantucket's growth as a whaling center. The locations of these lighthouses on the northern shore favored navigation toward the north bay around which Nantucket town grew. It was there that a wharf was built in 1723 by the general contractor and legendary Nantucket character Robert Macy, the grandfather of Obed Macy, who recorded his adventurous life:[19]

His practice was to bargain, to build a house, and finish it in every part, and find the materials. The boards and bricks he bought. The stones he collected on the common land, if they were rocks he would split them. The lime he made by burning shells. The timber he cut here on island. The latter part of his building, when timber was not so easily procured of the right dimensions, he went off-island and felled the trees and hewed the timber to proper dimensions. The principal part of the frames were of large oak timber, some of which may be seen at the present day. The iron work, the nails excepted, he generally wrought with his own hands. Thus being prepared, he built the house mostly himself.[20]

The growing number of houses in the town of Nantucket reflected its expanding population. Soon, the tryworks and furnaces that reduced whale blubber to oil were relocated from the cottages to the harbor, with a border of warehouses separating the smoky, smelly processing areas from the living quarters. Nantucket's relatively protected inner bay and investments made in infrastructure like lighthouses and wharves reinforced the town's status as the primary whaling settlement on the island, eclipsing Siasconset.[21]

The case of Nantucket and its whaling history demonstrate the nature of the dynamics between marine ecosystems and urbanization. The disputed economic concept of catch per unit effort (CPUE), a measure of how much energy is expended to capture a prey, is an important reference for understanding the complex relationship between land and sea. If a marine population is healthy, it is easy to land large quantities of fish quickly and enjoy a profitable fishery. But soon, increased fish mortality makes it necessary to expend more effort to maintain the profits, and fisherfolk must travel farther in larger boats, consume more fuel, and work longer hours. These measures are typical responses to a rising CPUE, which indicates that former populations have been depleted, and in this scenario, larger catches do not necessarily mean that there are plenty of animals. However, as a single indicator, the CPUE is an unreliable measure of biomass, as suggested by the account of the "greatest voyage ever made" published in the *Nantucket Inquirer* in September 1830.[22] It celebrated the return of *The Loper* with a load of 2,280 barrels of sperm oil harvested in a mere fourteen and a half months at sea. The achievement was exceptional at a time when most voyages took at least three years. A century after whaling was initiated, Nantucket invested in expanding its infrastructure, and although the business prospered, the need for longer voyages and larger crews cut into the profits.[23]

Marine historians work with ship's logbooks and other records to establish the impact of hunting on whale populations. Contributing to this collective effort, in 1935 Charles Townsend published four charts that map the occurrence of encounters and kills across the oceans.[24] The far-flung locations of these events can be explained by the habits of different whales, with feeding and mating grounds specific to each species. Townsend's charts have since been updated and their data consolidated, providing an accurate overview, if not of Nantucket's specific whaling habits, of American whaling as a whole in the eighteenth and nineteenth centuries.[25] The portrait that emerges shows the consistent positions of seasonal whaling grounds throughout the seas and confirms and explains the ever-increasing length of whaling voyages, a manifest sign of growing CPUE. This predatory pattern fueled the perception of whaling as a tireless chase of whales on constant retreat, reflected in literature by Ahab's pursuit of Moby Dick.

Inscribing Nantucket in the American whaling context, Eric Jay Dolin points out that the island's whaling industry ballooned following a shift in the main species targeted, which was also a shift to long-distance pelagic fishing.[26] Whereas New Jersey, Long Island, and other whaling communities were coastal fisheries that hunted right whales from the baleen family—which by the mid-eighteenth century were growing wary of dangerous coastal areas, if not less abundant—Nantucketers realized that sperm whales, from the Odontoceti family, were much more profitable, and armed their ships and crews accordingly. The major reason for this was the valuable spermaceti, but the sperm whale is also larger (adults can measure from 11 to 18 meters long and weigh

Siasconset, Nantucket, ca. 1860. Courtesy Nantucket Historical Association, Daintry Jensen Collection of Nantucket Photographs, PH54-63.

66

45 tons or more) with extraordinary thick blubber (up to 35 centimeters).[27] Unlike right whales, who frequent grounds close to shore, the largest Odontoceti make descents deep in the water column, feeding "mostly in waters of depths between about 300 and 1 200 meters."[28] This means that they are rarely found within thirty miles of the American coastline and thus were not accessible to open-boat shore fisheries. To catch sperm whales, Nantucketers needed to start "whaling in the deep" far offshore and to take on a faster, more powerful prey than the relatively slow and docile right whale. This was a bold move for Nantucket, a successful assertion of the island's leading position within American whaling. It also coincided with a major era shift in the whaling industry. Reeves and Smith's taxonomy characterizes the new approach as "American-Style Pelagic," a technique that grew from and eventually supplanted styles confined to the North Atlantic such as "American-Style Shore" whaling and "Basque-Style" traditions. Focusing the hunt on the sperm whale (even though other whales and marine mammals were also preyed upon) meant more than technical changes; it moved the hunt into the open sea, taking the whalers from the Atlantic to the Pacific and beyond.

It is difficult to prove that biological differences between right and sperm whales caused Nantucket's shift from coastal to pelagic whaling, but nonetheless the connection is tempting. What is clear is that the long-distance pelagic fishery affected the town's urban configuration, most importantly its harbor. As Obed Macy noted, the navigation of large ships into the harbor was impeded by a sandbar.[29] Several projects were developed to remedy the situation, the most inventive of which was the use of a system of floaters to lift the ships that was used from the 1850s to the 1870s.[30] The overall urban pattern of Nantucket did not change dramatically when the tryworks that had shaped the town's form were incorporated into the ships (as in the earlier Basque-style ships and in anticipation of the twentieth-century factory boat) because even though the ships now brought barreled blubber back to the island, further processing and refining was done on land. Overcoming the ecological limitations of shore fisheries stimulated industries that extracted and transformed whale products, and Nantucket grew along with them between the mid-eighteenth and the early nineteenth century. As the decreased catches and profits of coastal whaling in the first half of the eighteenth century demonstrates, without a change in the targeted species, American whaling would have had no future.[31]

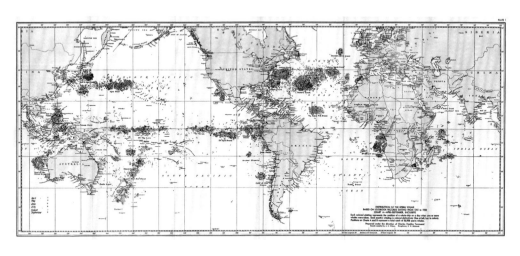

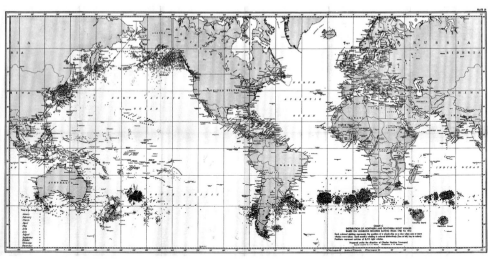

Distribution of sperm whales, April–September from 1761 to 1920 (above), and right whales, 1785–1913 (below), as shown by logbook records of American whaleships. Maps by Charles Haskins Townsend, 1935.

Sperm whales are unique animals.[32] One distinctive characteristic is their sexual behavior. Unlike other species, they are polyg-
ynous and travel in heterosexual pods comprising a patriarch, ten
to fifteen cows, and their young, often full-grown but not yet sexu-
ally mature.[33] Traveling at a safe distance behind the pod are often
two to three mature bulls "waiting to take over the cows when the
patriarch dies or becomes disabled," whereas other reproductive
bulls travel alone or in male pods.[34] After the breeding season, the
patriarch leaves to travel alone, and the pod becomes matriarchal
until the next season. Hunters prefer to target male sperm whales
as they are three to four times larger than females. So once a heterosexual pod is located and pursued by whalers, the typical mortality rate of males is significantly higher than that of females. The
effect on the pod's breeding capacity is negligible since females
will recruit a new patriarch when needed. An analysis of American
whaling logbooks suggests that killings were not indiscriminate
and that an approximate 42 percent of captures were large bulls,
a pattern that prevented major disturbances in the species' reproductive rate.[35] Additionally, the continuous hunt across grounds
"supports the argument that one ground was not hunted out
before whalemen moved on to another."[36]

Other whale populations were much more severely affected.
Such was the case of the North Pacific right whale, which in the
early 1840s was on the verge of being extirpated in a "remarkably
short and bloody chapter in the history of whaling."[37] After a consistent decline in captures and eventual exhaustion of right whales
in the northwest ground, whalers shifted their target to bowhead
whales in the Arctic, and then went on to hunt gray whales elsewhere. Hunting efforts like these, concentrated in space and time,
can have drastic consequences. Lance Davis, Robert Gallman,
and Karin Gleiter remarked that when whalers target the calving
grounds of monogamous species that typically form permanent
sexual attachments—and where the females are larger than the
males—they often kill pregnant or nursing cows, leaving behind
calves too young to feed themselves. In this way, entire populations
can be devastated very quickly. Some species were spared, however. For instance, the fast-moving humpback and the aggressive
gray whale—named "devilfish" because it can attack whaleboats—
were less appealing to American whalers because they produce less
and inferior oil and bone compared to other species.[38]

What was the architectural impact of such biological differences? As we have seen, hunting sperm whales on the high seas required a lot more fishing effort than inshore whaling, meaning larger boats, additional infrastructure on shore, and more time at sea. The average trip went from the early monthlong operations to voyages lasting up to four years. The hunt was sustainable for over a century because of its moderate impact on sperm whale populations and also because sperm whales were hard to capture. As a buffer to absorb the cost of the increased fishing effort between valuable sperm whale catches, whalers pursued less lucrative and more vulnerable species, such as the different right whale populations. Thus the biological differences between whale species made them complementary in an economic sense. And the exploitation of this complementarity affected the urban shape of Nantucket as it grew from a fragile coastal settlement into a highly specialized center of whaling. In contrast, the primary vocation of Siasconset—which continued to play a role in island networks and contributed workforce to the whaling effort—changed from a settlement built around shore whaling to a village hosting a shore-based cod fishery.

Men cutting first blanket piece from sperm whale and pulling blubber onto the deck. Photo: Clifford W. Ashley, 1904. Courtesy New Bedford Whaling Museum, 1974.3.1.35 and 1974.3.1.219.

As Obed Macy wrote of Nantucket's landscape in 1835, "the ground is not consecrated by deeds of chivalry: no ruined towers, no warlike mounds, no mouldering abbeys. … No spot is memorable for martial acts."[39] Yet Nantucket's main business still suffered from the chaotic effect war had on Atlantic navigation. For instance, a chart devised by Macy shows a 70 percent decrease in whale captures between 1812 and 1815, the years of the Anglo-American war.[40] One might assume that war between humans would allow the marine populations to grow in peace, but it didn't have much of an impact. As we have seen, specific populations of right whales—who have a long life cycle—had been so rapidly extirpated that a wartime pause would not bring them back. The sperm whales got just a short break from harassment. So despite Macy's point that the island was not touched by wartime combat, its development was nonetheless hampered by the disruption of whaling; at the same time, the conflict was relatively insignificant for the ecosystems.

The end of both the Anglo-American and the Napoleonic wars in 1815 changed the geopolitics of the Atlantic. A global increase in economic activities coincided with a mechanical revolution in which new machines made of steel and powered by steam replaced

Copyright 1903.
H.S. Hutchinson & Co.

older industrial equipment. Whale oil was in high demand as a lubricant for the Industrial Revolution and as a fuel to illuminate the streets of major cities, lighting up the nights of an urban society.[41] Between 1815 and 1861, these economic conditions favored Nantucket, and as a result one might expect to find the ultimate expression of whaling architecture there.[42] But that was not to be. The town boomed after 1815, building new warehouses, tryworks, commercial buildings, and wharves, as well as the houses and institutions that go along with them, but it was all razed in an epic fire on July 13, 1849. Witness accounts describe the flames being peppered by sudden explosions of the whale oil stored in warehouses and candle factories located around the city's harbor.[43] While a fire

Crew on the deck of the *California*, cutting up whale pieces for the tryworks, ca. 1903. Courtesy Nantucket Historical Association, Photographic Print Collection, PH165-P16559.

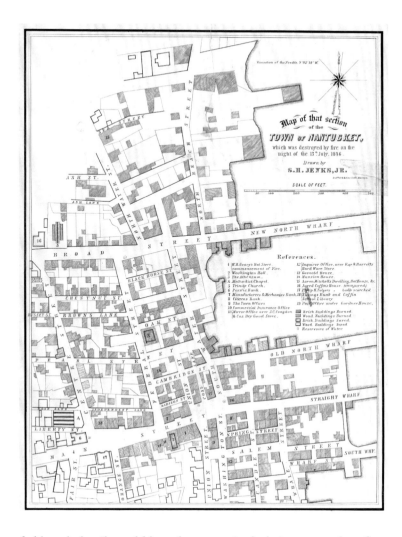

fed by whale oil would have been particularly intense, urban fires
were a common plague in settlements of all sizes, from the fires
that devastated the fishing village of St. John's in 1819, 1846, and
1892[44] to the infamous Chicago fire of 1871.[45] Nantucket's ability
to rebuild and recover its dominant position in the whaling business was hampered by a confluence of factors, including the longer
voyages required by the need to increase whaling effort. It became
difficult to recruit crews, especially at a time when the California
Gold Rush was drawing workers to the other coast.[46] More effort
meant higher operational costs, and financing whaling diverted
investment from urban reconstruction. There were also environmental limitations, including the scarcity of natural resources

72

(especially wood for new buildings and boats) that prohibited Nantucket from expanding its pressure on whale populations. So even as demand for whale products continued to grow, Nantucket was unable to meet it. And several sources concur that Nantucket whalers made a major mistake by insisting on chasing sperm whales in depleted whaling grounds while other captains sought new fishing grounds and diversified the targeted species.[47] Again, whale population factors might have factored into the island's urban success, or lack of it.

As Nantucket declined, New Bedford blossomed. Throughout the eighteenth century, this coastal town in Buzzards Bay had been laying the groundwork that allowed it to take full advantage of the expansion of the market for whale products in the nineteenth century. Maps of New Bedford from the 1830s display a series of wharves projecting from an extensive shoreline at the mouth of the Acushnet River.[48] It had established an important shipyard, was safe for deep-water navigation, and also benefited from the strong involvement of the Rotch family.[49] Joseph Rotch (1704–1784) and his son William (1732–1828) were major players in Nantucket's whaling industry, dominating every aspect of the business from the harpoon to the candle. In a canny maneuver to secure their empire, Joseph moved to New Bedford in the 1760s and set up business there. The Rotch family's knowledge and experience supported the development of a successful whaling industry in New Bedford, one that took over leadership of American whaling once the context no longer favored Nantucket. Although in a "sad condition" in 1812, New Bedford grew quickly, and by 1857 the town counted 329 working whaleships, approximately half of the American fleet.[50]

New Bedford's whaling landscape was different from Nantucket's. From the town's inception, it had a diversified economy, and its connection to a hinterland secured continuous flows of labor and capital.[51] While the business of whaling took place on and near the wharves, the town, with its orthogonal grid of streets lined with banks and insurance companies involved in the whaling business, projected a continental atmosphere in contrast with the informal settlements on Nantucket. Since many of the wharves were continuous with east-west streets, the large whaling ships and their masts extended the organized alignment of street façades into the bay. Descriptions of the city often mention two contrasting sectors: the busy, stinking wharves of Bethal Street, where every surface was impregnated with whale oil, and the wealthy and pastoral County

John Avery Parker House, County Street, New Bedford, 1832–1834. Photo: Joseph G. Tirrell, ca. 1895. Courtesy New Bedford Whaling Museum, 2000.100.85.100.

overleaf Street and waterfront with whaling ship docked at pier, New Bedford. Courtesy New Bedford Free Public Library.

Road neighborhood up the hill. The Greek Revival architecture of many of the city's institutions and prestigious homes is the work of Russell Warren (1783–1860), who was important in giving the burgeoning city a cosmopolitan flavor.[52] Among Warren's accomplishments was the Pearl Street Train Depot, a railway terminal that connected the harbor city to the continental rail system. While modest in size and material, it was built in an avant-garde Egyptian style that would later become popular in many architectural circles.[53] As of its inauguration on July 1, 1840, the terminal was a symbol of the important link between New Bedford and Taunton, which since 1836 had been served by a branch of the Boston-Providence railroad running from Mansfield.[54] The railroad gave New Bedford's economy a tremendous boost, making the city a hub connecting deep-water navigation and hinterland distribution networks. This significant infrastructural advantage, combined with the local expertise in hunting and marketing, allowed New Bedford to quickly eclipse Nantucket as America's leading whaling port.

New Bedford's success, its urban landscape, and its growth patterns were shaped less by fishing than by an emerging industrial society. As Lance Davis, Robert Gallman, and Karin Gleiter have carefully pointed out, the market for whale products floundered even before the discovery of oil wells in Pennsylvania.[55] It was a matter of competition: alternatives to whale oil such as manufactured gas, lard, and tallow oil claimed growing shares of the American market after 1830. Initially, the negative effect of competition on the whaling industry was compensated by overall growth in the fuel sector and increased exports. Prices were high so the profit margin of whaling voyages increased, but at the same time productivity was dropping due to longer voyages and the technical limitations of whaling.[56]

A sign of New Bedford's shift from whaling to a diversified industrial economy was the operations of the New Bedford Cordage Company. Organized in 1842 with significant capital from the Rotch family, the company imported manila fiber from the Philippines to supply rigging rope to the whaling fleet.[57] The cordage was one of many businesses on New Bedford's wharves, a lively zone where outfitting, repair, cargo handling, and storage all took place. It was populated by blacksmith and artisan workshops, warehouses and storerooms, sail lofts, chandleries, and counting houses that complemented the candle works and oil processing facilities built within the urban fabric close to the harbor.

New Bedford's waterfront was a landscape shaped by barrels and the circular forms of tanks and refining kettles. A detailed plan of the area drawn up by Mark Foster identifies over twenty-seven oil refineries active between 1800 and 1860.[58] The factories developed in successive expansions consisting of masonry buildings one or two stories tall with pitched roofs and glazed skylights that illuminated the bleaching vats. The complexes were peppered with smokestacks and arranged to form courtyards that were filled with barrels. A refinery building usually had brick peripheral walls that enclosed a wooden internal structure. At its core was a hearth, also of brick, that provided the heat needed to refine the oil. The main hall contained large wooden presses with levers that blended in with the building's beams and trusses. Surviving photographs show surfaces saturated with leaked grease and a geometry of structural elements in combination with white stacks of brick-shaped spermaceti cakes. They also show two modes of transportation: the horses used to haul cargo and the omnipresent parallel lines of railway tracks. The latter serve as a reminder of how the rail link allowed for the continuous movement of whale products toward the hinterland markets, an infrastructural advantage New Bedford enjoyed over the island of Nantucket.

While the most prominent buildings at refineries were candle houses, the bleaching vats stand out as whaling architecture's most intriguing spaces. As banal places of production, they did not represent the new kind of building that emerged because of whales. Yet, they fueled the industry's sense of identity as the site where the final product was whitened. Sunlight and air make the oil's impurities sink in shallow vats, creating a lighter color and a superior product for the trade. There is little information about these spaces, which vanished over time. Above the rectangular lead-lined vats, the rational repetition of building structural elements dialogued with the pipes and drains used to pump and collect the oil. They were located in the loft spaces of the refineries, where glass-tiled gabled roofs flooded the interior with a luminescent atmosphere, infusing the images that have survived with a visceral pathos. These spaces are imbued with a mysterious anachronism, the geopolitics of the oceans and the planetary breath of whales scaled down and housed in an improvised greenhouse. While the inner life of the leviathan was a matter of theological imagination, the architecture of whaling involved prosaic constructions.

Although historians agree that an increased demand for whale oil during the first half of the nineteenth century came from New England's textile industry, the effect of cotton mills on an urban landscape attributed almost exclusively to whaling has not gotten much attention.[59] But as Kingston Heath pointed out, the cotton industry had a significant share in shaping New Bedford.[60] Although it was not the first city in the area to develop a textile industry—the neighboring cities of Pawtucket and Fall River can claim precedence—New Bedford's cotton mills benefited from the investment of surpluses accumulated during the peak years of whaling. A drawing signed by engineer Seth H. Ingalls (1806–1884), superintendent of construction for the local railroad, shows the arrival of the Taunton railroad to Pearl Street and an additional branch connecting it to the Wamsutta Mills, a cotton weaving factory established in 1846.[61] The drawing contrasts the orthogonal geometry of a new wharf and contemplated streets with the irregular lines of existing points, canals, and pounds along the river. Assessing the ecological effects of New Bedford's urbanization, a research group determined

78

that by 1851 whaling wharves had covered 15 hectares of the estuary's surface, probably causing "the direct destruction of benthic habitats" and reducing "the volume of water passing through" the canal between New Bedford and Fishing Island "during tidal exchanges."[62] As the number of mills increased—from two in 1846 to thirty-one by the 1920s—so did the production of industrial waste, which had a major impact on the local ecosystem. Even more harm was caused by the physical growth of the city and its population as the mills expanded and attracted a large immigrant workforce. New Bedford's characteristic three-decker architecture was developed to provide housing for these textile workers who lived, worked, and generated sewage in new developments over the wetlands north and south of the original settlement that modified, damaged, and eventually destroyed "essential functional ecological attributes of the Acushnet River estuary."[63]

The demise of American whaling occurred at the intersection between the decrease of oceanic natural resources and continental

Rotch's North Wharf, New Bedford, 1876. Photo: Joseph G. Tirrell. Courtesy New Bedford Whaling Museum, 2000.100.360.

Robinson & Co. Oil Works, refined spermaceti in bulk and cakes, New Bedford, 1927. Photo: William H. Tripp. Courtesy New Bedford Whaling Museum, 2000.101.64.2.

urban growth. While initially whales could be sighted from the shore, hunting pushed them ever farther away. On Nantucket, islanders responded to market fluctuations by investing in minor technical adjustments and increasing their fishing effort by making longer voyages. The relation between the fish and the shore was obvious: as the whales moved farther and farther away, so did the whalers. Nantucket succeeded Siasconset with ingenious methods for improving harbor navigation, but its building practices and urban patterns remained preindustrial. In an 1811 letter, the traveler Joseph Samson described Nantucket with candor:

> Such is the simplicity of this primitive place ... that the streets which have branched out from each other by imperceptible degrees, every man being at liberty to place his house according to his own fancy, and being naturally more disposed to regulate his front by a point of compass, than by the direction of the street.[64]

After 1815, during its "Golden Age," the reality of the whaling business and its markets shifted, and natural resources became scarcer. Competing in this climate required new market strategies, and with Nantucket's preindustrial configuration and environmental limitations, it was unable to meet the industrial demand for oil products. Successful operations required large factories

and refineries and efficient connections to distribution networks, both of which require investment to set up. The exhausted whale populations could not bear the extra cost of insularity. When the continental port of New Bedford absorbed the whaling business, its backers were wise enough to diversify, adding cordage companies and textile mills to their investment portfolios. The urban landscape of New Bedford might well represent the final type of an American whaling town, but it also marks the final stretch of the American whaling effort.

Tragedies are telling. The 1871 disaster that struck a forty-ship whaling fleet off Point Barrow, the northernmost headland in Alaska, is a powerful evocation of the vast distance between the whales and their urban counterpart. The Arctic fleet began to pursue bowheads north of the Bering Strait in 1848.[65] At first, the effort was fruitful, but captains soon questioned the wisdom of hunting that far north in cold, rough waters. One New Bedford captain wrote in 1852, "I felt as I gazed upon the great frozen icefields stretching far down to the horizon that those were barriers placed there by Him to rebuke our anxious and overweening pursuit of wealth."[66] In 1871, there was no doubt that it had been a bad idea. Very late in the season at the end of August, a hostile southwest wind prevented all but seven ships from sailing south and simultaneously pushed the pack ice against the shore. Come September, massive ice floes from offshore began to crush ships against

Robinson & Co. Oil Works, spermaceti from taut press to be further refined, with refining kettles, New Bedford, 1927. Photo: William H. Tripp. New Bedford Whaling Museum, 2000.101.64.3.

81

the ground ice. After one ship was broken up "like an egg-shell," the fisherfolk regrouped and continued whaling, but by the time a third ship was wrecked on September 8, the captains realized that they were trapped.[67] They knew that the seven ships that had escaped the floes were still just south of Icy Cape, so on September 14 they began an exodus of whaleboats through fierce gales, white-capped waves, and freezing winds. It took the 1,219 passengers and crew two days to safely reach their rescuers, but in the evacuation thirty-three ships were abandoned and lost, twenty-two of them from New Bedford. The fishing effort of the golden era of American whaling era had reached its limit.

Nonetheless, it did not end there. New Bedford whalers kept on despite repeated tragedies, but the city did not depend on whales for its urban growth and survival. New Bedford eventually became a textile center, whereas Nantucket became a summer resort. Even today, both places market themselves by parading their whaling histories as a glorious formative past.[68] Unfortunately for the whales, new fishing techniques—including Norway's steam-powered ships armed with bow cannons to fire harpoons with explosive heads—were developed and used to supply different markets and support the production of different types of whale products such as baleen corsets and whale meat. The American pursuit of oil had a negative impact on marine ecosystems, but things would get much worse.

Robinson & Co. Oil Works, vats for oil bleaching by exposure to the sun, New Bedford, 1927. Photo: William H. Tripp. Courtesy New Bedford Whaling Museum, 2000.101.64.6.

overleaf Arctic Ocean, the *Bear* getting free from the ice pack, 1898.
Courtesy New Bedford Whaling Museum, 2000.100.200.74.

3 *The Harbor and the Factory*

The French word *criée*—from the verb *crier* whose Latin root is *critare*, to yell, shout, or cry for—designates a building where fish auctions take place. It is a crucial interface between land and sea, one where a natural resource becomes a commodity and where the fortunes, or misfortunes, of fishing enterprises are made. As the name suggests, the cacophony of auctioneers, buyers, fisherfolk, investors, carriers, inspectors, and voyeurs engaged in interrelated tasks creates a hectic and noisy environment at odds with the rhythmic architecture of many of France's *criées*.

The Halle des Mareyeurs in Lorient, a major fishing harbor in south Brittany, is a case in point. It is a symmetrical 235- by 38-meter structure with a central cross axis. On the long wharf façade, parallel to the water, is a deep open-air shed, and behind that is a two-story interior space. It was completed in July 1927 as part of the development of Kéroman, a marshy site south of the older harbor facilities near Lorient's city center, into a modern port. The aerial photographs that record the opening of the Halle des Mareyeurs are telling of the massive transformation of the estuarine ecosystem required to build the new harbor and of the auction hall's proximity to the railroad. The long building is a buffer between the boat and the railway, a key logistical element within the supply chain.

Inauguration of the auction market in Kéroman fishing harbor, Lorient, 1927.
Courtesy Archives Municipales Lorient, 7Fi694.

A contrasting example is the 1935 competition project for the Trouville auction hall in Normandy, won by Eugène-Maurice Vincent (1887–1956) and completed in 1937 in collaboration with Maurice Halley and Marcel Davy.[69] At approximately 26 by 10 meters, it is about ten times smaller than its predecessor in Lorient. It sits in the heart of Trouville between the wharf, the city hall, and the casino, its regionalist architectural language blending in with the rest of the summer resort town. Supported on piles that straighten the bank of the river facing Deauville, the concrete structure is clad in brick plastered to give the walls the appearance of stone masonry and oak framing, the whole covered with wooden pitched roofs and a half-timbered gable emphasizing the "Norman" architectural flavor.[70] The streetside façade is equipped with booths for retailers, and the central auction hall is covered with a barrel vault.

The contrasting architecture of the *criées* of Trouville and Lorient suggests a major difference in fishing operations, the former often characterized as artisanal and the latter industrial. In Trouville, not only was the half-timbered "Norman" style considered appropriate to the urban setting, but it also reflected the small-scale, family-based operation of the coastal fishery, whose clients came down to the quays. Lorient's rough-looking modern facility was designed for an industrial fishery that served markets in the main urban centers by train. While distinguishing between artisanal and industrial practices can obscure the overlapping knowledge, techniques, and procedures they share in terms of both fishing and architecture, the binary is useful in understanding the complex dynamics unleashed by the industrialization of fishing practices in Brittany, which led to the transformation of its landscape and its marine ecosystem.

Artisanal fisheries are typically described as simple or small. Definitions emphasize the household-based and local nature of the business, with the fishers being self-employed or closely related to the fishmonger. Despite the connotations of the name, "artisanal" fisheries are not necessarily low-tech, nor are the catches always consumed locally. Any non-artisanal operation is by default "industrial," a term that implies complex operational systems, large investments of capital, and fisherfolk as employees. Various regulatory agencies distinguish between small- and large-scale operations, and although some go by the dimensions of the fishing vessels, the clearest feature of a small or artisanal business

is that it is family-based, with the owners directly involved in the
fishing activity. Nonetheless, the distinction has been blurry for
a long time. A much-studied historic example of how investment
capital interferes with local exploitation of marine ecosystems
dates from Newfoundland's colonial period.[71] The perception of
a gap between artisanal and industrial fisheries was boosted in
the nineteenth century when the introduction of technical innovations resulted in the artisanal frequently being relegated to the
realm of "tradition," if not of "superstition," whereas the industrial was referred to as "modern" and "progressive." In reality the
differences were more subtle and the connections deeper than
this language suggests.

Sardine fisheries in nineteenth-century Brittany were coastal
operations.[72] In the summer, fish was found just a few miles off
the coast and pursued from small *chaloupes* using roe as bait and
drift nets.[73] Boats would land their catches in one of a number of
natural harbors, bays, and beaches where the fish would be processed in a small-scale facility. Before the 1820s, they were salted
and pressed in barrels, and later increasingly canned in oil. In both
cases, despite the complementarity between fishing and industrial processing, the fisheries can still be described as artisanal
since the processing was performed by a separate business. As
we have already learned, the fatty acids of *Sardina pilchardus* make
the fish rot rapidly after death, making processing essential to

preserving the product and securing its commercial value. And thus the factory enters the scene.

Duhamel du Monceau's 1769 treatise on "fisheries and the history of the fish they supply" describes the minutiae of making pressed sardines.[74] The engravings depict a busy shoreline with many people carrying trays loaded with fish from the waterfront to workshops, where they would be smoked or salted. The salted fish were then mechanically pressed into barrels by means of levers and other devices. Every step in this proto-industrial production process required human energy. Xavier Dubois's thorough study of the sardine fisheries of southern Brittany in the nineteenth century describes how the small pressing workshops of the 1830s were inserted within the "urban" setting of coastal communities, close to the docks and the road, well connected to commercial routes.[75] Such installations were operated by businessmen who used Norwegian pressing techniques and who regularly imported roe to use as bait. Duhamel's plates do not show the merchants or the presses, making it hard to visualize the architectural character of the industry, but the buildings seem to have followed conventional building practices for warehouses and workshops.

An industrial breakthrough came in 1824, when Pierre-Joseph Colin (1785–1848) turned his father's workshop near the harbor in Nantes, on the Loire estuary, into an efficient cannery. This transformation was part of a structured agro-industrial development

Fig. 1.

A

Fig. 2.

A

Fig. 3.

A

Angel.Motte Sculp.

of the region between Nantes and Bordeaux led by former slave traders following the 1815 abolition of the slave trade. Colin's cannery did not make him a national hero like Nicholas Appert (1749–1841), who patented the canning process, but it was a successful venture.[76] Colin experimented with preserving food in tin cans. He discovered that sardines were tastier once fried and packed in oil, an ingredient that prevented the metal can from corroding and contaminating the food that had doomed other canned products.[77]

facing Sardine fisheries and fishing gear, pressing workshops, in Duhamel du Monceau, *Traité général des pêches et histoire des poissons*, vol. 2, section 3, plate 18, 1772. Courtesy Bibliothèque Nationale de France.

Brittany's coastal sardine fisheries were seasonal, active from midspring to late summer, and just like the complementary fish yields throughout fall and winter, fresh fish were more valuable than processed ones.[78] Sardine catches are notoriously irregular because sardines are forager fish whose populations are affected dramatically by fluctuations in climatic and marine conditions.[79] During years of plenty, when prices drop, revenue was still good because press operators could buy at a lower price and preserve more fish. The abundance of sardines in 1822 seems to have played an important part in the canning breakthrough.[80] The low prices that year meant that Colin could engage in the trial-and-error experiments necessary to develop a tasty canning recipe without creating expensive waste along the way. The successful result upgraded sardines from a brined low-cost commodity to a canned high-end product, generating demand from a new clientele.

While Colin's success came quickly, it took time to transform the industry. Markets had to expand and consolidate, processes had to be adjusted and replicated by competitors, and production techniques had to be perfected, which happened in 1852 with the adoption of the autoclave to properly sterilize cans. Throughout the 1840s and 1850s, Brittany's canning industry grew steadily from a few canneries in Nantes to operations distributed throughout the coastal villages of Finistère and Morbihan, eventually extending all the way from Brest to La Rochelle.[81] Most of the product was exported using trading networks that were already in place, which might explain why, of the many countries that had access to sardines, France became the world leader in exports of the canned product, and why sardines were synonymous with French cuisine. Unlike other fish, sardines were not sold at auction but traded directly between the fishers and the factory. This direct relation between canneries and fisherfolk played a pivotal role in the growth of sardine fisheries. During the second half of the nineteenth century, retaining walls, slipways, wharves, and piers were built to facilitate efficient transfer of catch from the boats to the processing facility. Although these infrastructures tended to be built over the commons, they were paid for not by the state or municipality but by the factories and their investors.[82] The improved conditions would attract fisherfolk, and the quicker transfer of the fish would prevent decomposition and ensure a better final product.

The growth of the food-processing industry prompted the expansion of Brittany's sardine fisheries. The production of barrels of pressed sardines, a proto-industrial activity that provided a low-price commodity for national consumption, evolved into an industry focused on the export of high-priced cans. This shift was made possible through local and regional capital investments, with competing factories spread throughout coastal communities.[83] Port-Louis, Étel, Quiberon, and the islands of Groix and Belle-Île-en-Mer all experienced an increase in demand for sardines. This came from traders and industrialists, who were able to retain an upper hand over fisherfolk because they had access to credit and thus could more easily own the key elements of fishing, like imported roe and boats. The sprawl of presses and canning factories in the small centers along the coast benefited from the dynamic cabotage trade, which allowed fisherfolk to operate over a large portion of the Bay of Biscay. The growing international demand for canned sardines, which led to this flurry of fishing and processing activity, peaked in the early 1870s, driven by events such as the California Gold Rush and the American Civil War between 1861 and 1865.

In a canning factory, raw materials are transformed into a final product in various steps. First, the sardines are washed and beheaded, which produces a large volume of sewage and remnants that can be used to make by-products like fish oil and fish meal. The sardines are then cooked and canned. In parallel, tin cans are produced in metal workshops, and often a printing press runs in tandem to create labels that give the product its commercial appearance. Many canneries were upgrades of older pressing plants that were repurposed and enlarged over time, with more personnel employed in a growing number of tasks. The mechanization of the production chain is reflected in the cannery architecture, which typically includes not only a central canning hall but also offices for managing administrative tasks, storage warehouses, and the factory chimney, which became an iconic feature of canning landscapes.[84]

In bathymetric plans of the Bay of Biscay, the deep waters of the southern part of the bay contrast with the north, where the extension of the continental platform results in a relatively shallow area hospitable to sardine populations.[85] Their seasonal movements are hard to trace, and spawning grounds vary in relation to currents, water temperatures, and plankton availability.[86] Even so, research by marine biologists Suzanne Arbault and Nicole Lacroix in the 1970s revealed a rather consistent seasonal distribution of

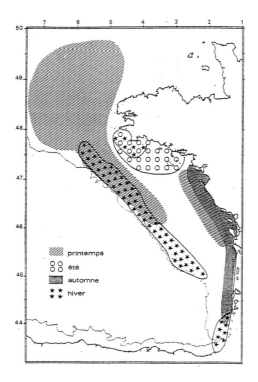

the species:[87] in the fall, near the southwestern part of the French shore between La Rochelle and Biarritz; in the winter, on the southern ridge of the continental platform; in the spring, moving northwest toward the English Channel; and in the summer, concentrated near southern Brittany.[88] Fluctuation within such patterns helps to understand the unpredictability of catches and the frequent "bad years" experienced by sardine fisheries.

Extreme variations in sardine populations are not exclusive to Brittany and affect foraging species since they are particularly sensitive to environmental factors. Sardine populations feeding the canning factories of Monterey Bay in California, an urban area

Distribution of sardine spawning grounds in different seasons, according to Arbault and Lacroix, Gulf of Biscay, 1977.

developed in the early twentieth century and made famous by John Steinbeck's novel *Cannery Row*, are a case in point.[89] Paleo-ecological research in the Monterey Bay area has revealed intricate patterns in how sardine populations rise and fall in relation to water temperature and other species of fish.[90] Climate change and climatic phenomena provoke changes in water temperature, which affect the amount of phytoplankton and zooplankton available for sardines, and anchovies, to eat. When the population of Pacific sardines living off Monterey Bay collapsed, the anchovy biomass swelled to fill the leftover ecological space, and scientists speculated that sardines could be restored by capturing the anchovies. The idea was never tested, and after a few years, the California sardine population recovered without any obvious explanation. Such fluctuations in sardine biomass, known as "regime shifts," are globally synchronous, which attests to the impact of climate on fish population dynamics.[91]

Similar patterns, with specific seasons proving disappointing in terms of catches, were recorded in Brittany as early as 1831, when barrel-pressed sardines were the main output of fish landings.[92] At that time, the economic impact of regime shift fluctuations was mitigated by factors such as the specialization of the sardine business, its seasonal character, and its complementarity with other sources of income. However, the story was quite different after the introduction of industrial canning. Because of the number of people directly involved in the business and dependent economic activities, the "bad years" of the late nineteenth century became tragic episodes of famine and despair for many Breton fisherfolk.[93] By the 1870s, the novel science of oceanography began to look for causes, patterns, and solutions to the consequences of sardine population fluctuations.[94] Research identified climatic conditions, increased trawling, and the quality of bait as factors. Some were even concerned that religious missteps were to blame.[95] Although overfishing was not usually mentioned, fisherfolk were skeptical and did not invest in technology to overcome the decrease of resources, retaining the traditional gill nets instead of adopting seining. In this way, the sardine fisheries weathered the ups and downs for decades, with highly profitable years, such as 1881 and 1882, used to offset periods of disaster.

These disasters may have been triggered by regime shifts and the variable nature of sardine biology, but they were exacerbated by other issues. As Xavier Dubois points out, contextual factors turned population issues into crises. For instance, in 1890, 1894, and 1895, most of the sardines landed were too small to be canned.[96] Not only were the fish wasted, but there were fewer specimens available the following season. The standardized industrial can was not flexible enough to cope with natural cycles. As a result, it was the canning industry itself that had to change to accommodate the unpredictability of sardine supply. The number of installations in the Morbihan area was almost halved in 1891, reduced to thirty-two from the sixty that were active in 1879.[97] In its focus on producing a high-end export product, Brittany's canning industry neglected the French market, and, after demand peaked during the American Civil War, international competition took control of various markets by providing less expensive products. In many ways, the sardine crisis was brought on by competition and the coastal—if not artisanal—fisheries of Brittany could not keep up. Not only were they struggling to provide fish, but the associated costs they had to bear were higher than those in competing countries. Between the price of olive oil, which was higher in France than in Portugal and Spain; the 1881 ban on lead solder to seal cans, which required paying more for tin pewter;[98] and increased exploration costs due to the vulnerability of resources, Brittany could not supply fish at the volume and price needed to compete in a globalized market.

In this context, the fishery of the island of Groix adapted by changing its targeted species from sardine to albacore tuna (*Thunnus alalunga*), an apex predator with flesh that can be canned in a similar process to that used for sardines. Fishing for tuna, however, is done on the open sea and requires larger boats and the harbor infrastructure to dock them. This adaptation was possible for an island like Groix but more challenging for the fishing communities of continental Brittany, whose small sardine boats were not suited for offshore navigation. Most fisherfolk could not afford a larger boat—and even if they could, vessel sizes were limited by the infrastructure of the small fishing harbors that dotted the coast.[99]

Ecological issues alone did not cause Brittany's sardine crisis; it also resulted from the growing dominance of canning factories and the limited capacity of regional infrastructure to support them. In sum, artisanal fisheries that gave rise to the industrial production of sardines were expected to industrialize but lacked the capacity to do so. Furthermore, the insecurity of sardines as a resource made them too risky to attract capital investment.

The discrepancy between the artisanal fishing of sardines and the related industrial processing, a consequence of the demographics of the targeted species, fostered a major architectural transformation in which the harbor replaced the factory. In this process, fisherfolk were driven away from fisheries in a split that was motivated by capital. Most of the literature depicts Breton sardine fisherfolk as "conservative" for rejecting novel (and expensive) techniques to improve catches like purse seining, which was adopted by various pelagic Atlantic fisheries. At that time, in the late nineteenth century, trawling (a method not suitable for sardines) was gaining in popularity. With the benefit of hindsight, we can observe that the capital invested in sardine fisheries was mainly local, whereas the majority of investors in trawling businesses were companies located in Paris or other French regions. Trawling, while it could benefit from some of the local knowledge and infrastructure associated with the sardine fishery, was an entirely different business: it targeted other species, with different habitats and habits, in new locations that fostered new ways of exploiting marine resources. Trawlers went after plaice, for the most part, as well as sole, turbot, and other benthic fishes—species living at the bottom of the water column. And as we know, profits can be high when new fishing grounds are first explored, and high profits attract capital. In late-nineteenth-century Brittany, the industrial trawling fisheries, associated with harbors, started to overlap—without necessarily competing—with the artisanal sardine fisheries associated with canning factories.

Architects claim that capital flows through the built environment. This was the case for the development of harbors to support trawling and the exploitation of benthic ecosystems in the second half of the nineteenth century. Investment in new infrastructure, motivated by changes in consumer behavior, transformed harbors into interfaces between fishing technologies and consumer markets. The sea-land continuum typical of coastal fisheries, where fisherfolk landed their catch directly into the food-processing factory, was broken down into multiple steps connecting the ecosystem with the consumer. Complex supply chains developed, connecting primary production with households and caterers through commercial networks that linked production, auction and processing, distribution, retail sale, and consumers in a series of steps.[100] This sequence profited from important technological transformations within the various links in the chain. First, boats

100

were mechanized. Steam and diesel replaced sails, motorized capstans and hoists replaced arms, large trawls replaced nets. Then, ice and refrigeration were used to prolong the shelf life of captured fish. And, most importantly, trains sped up, efficiently delivering fresh fish from the shore to the city. Ultimately, the most profitable and secure fishing business was no longer the factory by the sea but the harbor directly connecting the marine ecosystem to the consumer.

Although this process is evident in the history of fishing infrastructure in Brittany and elsewhere in France, its developers followed the example of English industrialization. Among multiple factors that led to the growth and diversification of fish intake in nineteenth-century England, the railway connection between Manchester and Hull was a key trigger since it connected the ballooning populations of industrial cities to fresh food from marine ecosystems, mostly in the North Sea. The historian Robb Robinson, who traced the expansion of British trawling, points out how the cost of freight carriage prevented city merchants, railway companies, and fisherfolk from seeing the train's potential

Port-Haliguen, Quiberon, 1902. Photo: Gabriel Ruprich-Robert. Courtesy Ministère de la Culture (France), Médiathèque du patrimoine et de la photographie, AP60L01442.

overleaf Sardine nets in commercial basin, Lorient, ca. 1906. Photo: Henry Laurent. Courtesy Archives Municipales Lorient, 5Fi1542.

PERSÉVÉRANCE

DIEU ET PATR

to grow the fish trade.[101] Fresh fish was perceived and negotiated as an upmarket commodity in comparison to cured or processed fish, so it was not in wide demand. But when train companies lowered the freight cost, fresh fish sped along the rails into urban markets, becoming an accessible protein for the working class which eventually led to changes in consumption habits and associated market growth.

Before the railway, the River Thames brought fresh fish into London markets via ice boats. Fish spoil quickly due to enzymes and bacteria, a process that is slowed by chilling. In England, the fishing fleets of Harwich began using ice to preserve fish at the end of the eighteenth century.[102] The ice, collected in winter and stored in icehouses for several months, was instrumental in securing the time needed to bring fresh fish to urban markets, of which London was the most important.[103] As London's urban population swelled, so-called "wet" fish became a widely available, but expensive, alternative to cured fish. In fact, the capital was unique among British inland cities in its proportion of fishmongers to butchers, in part due to its size and centrality. In 1842, London had one fish dealer for every four butchers, whereas in Warwickshire the ratio was one to twenty-seven, in Staffordshire only one to forty-four.[104] These relations would soon shift, propelled by the combined effect of urban growth, railways, and trawling.

The first architectural sign of the prominence of fish in London was the Fishmongers' Company Hall. The Fishmongers' Company has administered the city's fish trade since the thirteenth century, and with powers reinforced by a 1698 Act of Parliament, it secured the concentration of the trade in Billingsgate between the river and Lower Thames Street around Fish Street Hill. Its classical headquarters, just west of London Bridge, is worlds away from the frailty and vernacular character of most fishing facilities. Designed and built between 1831 and 1834 by Henry Roberts (1803–1876), at a time when George Gilbert Scott (1811–1878) was an apprentice in his office, the Hall's monumentality inscribes it within a network of reputable urban institutions.[105] Its dignified character does not connect it to fish or to the ecological and environmental contexts of fishing but rather reflects the social aspirations of the fishmongers. Their economic power, already visible in the 1830s before the

Billingsgate market from Lower Thames Street, London, ca. 1900. Photo: Keasbury-Gordon. Courtesy The Keasbury-Gordon Photograph Archive, Alamy DABM8X.

Railway Mania of the 1840s, came from the growing consumption of fish products within the industrial city.

Billingsgate was a loud, frenzied area flavored by the odor of fish, with barges docked at piers that were extensions of city streets. Before the 1850s, when a proper market hall designed by James Bunning (1802–1863) was erected, fish were sold from a "group of sheds and stalls, accumulated no one knew how," to use the words of a contemporary.[106] The concentration of the city's fish trade in a single location puzzled many contemporary chroniclers. But that concentration made sense given that the product was fresh fish. Fish supply is more unpredictable than other food items like meat and vegetables because of the nature of fisheries and their dependence on the ecological dynamics of marine environments. Variables such as seasonal trends, meteorological and navigational conditions, and the behavior of fish made the fresh fish trade a highly specialized niche with volatile prices. As a result, buying at the wholesale market—which adopted the frantic Dutch auction system, where the lot is sold to the first bidder at the highest price, instead of the last going up—required expert knowledge that discouraged direct trade between individuals or households and sellers. Instead, fish bought wholesale at Billingsgate was distributed to retailers in neighborhoods throughout the metropolis.

Market halls of the nineteenth century like Billingsgate's replaced outdoor market squares. As architectural historian Nikolaus Pevsner (1902–1983) noted, the new buildings were a curious hybrid of the railway station and the exhibition hall of an international fair.[107] The central covered space was surrounded by stalls and galleries that made up the urban façade of the market and mediated between the street and the interior. An important innovation was the cast-iron roof structure developed in 1835 for Hungerford Market by Charles Fowler (1791–1867), who also designed the renovation of Covent Garden.[108] Hungerford Market was one of various unsuccessful attempts to circumvent Billingsgate's monopoly of the fish trade that likely failed because Billingsgate's riverside location was better suited to the unique needs of fish supply.[109] Ironically, the 1850s design for Billingsgate Market was a successful adaptation of Fowler's type, as its longevity suggests: after being rebuilt in the 1870s, it remained open until 1982, when the wholesale market was moved east to Canary Wharf and the historic market building was converted to an office space by Richard Rogers (1933–2021).

In 1842, Charles Knight (1791–1873) wrote an atmospheric description of Billingsgate, dominated by the presence of fog and boats moored on the Thames:[110] "The boats lie considerably below the level of the market, and the descent is by several ladders to a floating wharf, which rises and falls with the tide, and is therefore always on the same level as the boats."[111] Fish would soon arrive by rail in increasing numbers.[112] The statistics record 108 tons of water-borne fish in 1848 and 95 tons water-borne plus 71 tons rail-hauled in 1875. By 1910, the year the total tonnage reached its peak, 128 tons were rail-hauled and only 70 were water-borne. The tonnage of water-borne fish continued to drop between 1910 and 1936, when the supplying fleet was dismantled. But already in 1842 Knight insisted on the effects of railways providing fish supplies, not only to London but also to regions of England where fresh fish was not previously available: "In Leeds, Manchester, and Birmingham, during the summer of 1842, the supplies of fish, chiefly by the railways, were occasionally immense."[113]

One example of this new accessibility in 1842 was the Manchester shop opened by the Flamborough and Filey Bay Fishing Company. With the support of low transport rates and fast steam-powered trains, the shop would sell in the afternoon, at a competitive cost, fish landed overnight in the harbor of Hull. The low

Billingsgate market weathervane, London, 1931.

107

prices and novel product made the shop an immediate success, with demand exceeding supply and customers forming long queues. Numbers soared, the 3.5 tons of fish transported per week by the Manchester & Leeds line in 1841 grew to 80 tons in 1844, and other fishmongers and suppliers increased the flow of fish through the railway to the steadily increasing urban population.[114] With the help of the extensive British railway system and the telegraph that allowed traders to ascertain the best deals throughout the distribution network, markets were connected to different fisheries, and London became the center of fish consumption, offering a combination of species harvested in various geographic locations.[115] The historical southern harbors of the Channel, where sail trawling had been used for a long time, began to see competition from the new trawling fleets of the North Sea. The competition meant more supply to meet increasing demand from the growing urban population.

The first Billingsgate market building opened in August 1852.[116] Designed by city architect James Bunning, it was an ingenious

Horace Jones, Billingsgate Market, London, 1877. Lordprice Collection/ Alamy B77BPE.

machine.[117] First, a retaining wall was erected by the river, forming a terrace that expanded the available commercial surface. Public access was from Thames Street and led to the upper floor; the lower floor was connected to the riverside. George Dood describes the roof as "formed of semi-circular bays of corrugated iron, supported by iron pillars, and lighted by thick rough glass." The market's most prominent feature was the clock tower facing the river, the main purpose of which was ventilation. The mechanical system designed by Henry Bessemer (1813–1898) used "revolving pump-discs" to drain "the air from the lower vaults and market by suction, and propel it up the air-shaft enclosed within the clock-tower; by which contrivance it is said that 50,000 cubic feet of contaminated air can be expelled in a minute." The water needed "for washing fish, washing hands, washing the market, and keeping everything neat and clean" was pumped directly from the Thames.[118]

Describing the London "of the poor" in 1851, Henry Mayhew provided a living portrait of the fish trade at Billingsgate. He emphasizes the role of railways, "by means of which fish can be

James Bunstone Bunning, Billingsgate Market, London, 1851. London Metropolitan Archives (City of London)/Heritage Images, Alamy DDMPWF.

sent from the coast to the capital with much greater rapidity, and therefore be received much fresher than was formerly the case" and the resulting transformation of fresh fish from an expensive product to a low-cost commodity:

> This cheap food, through the agency of the costermongers, is conveyed to every poor man's door, both in the thickly-crowded streets where the poor reside—a family at least in a room. ... For all low-priced fish the poor are the costermongers' best customers, and a fish diet seems becoming almost as common among the ill-paid classes of London, as is a potato diet among the peasants of Ireland. Indeed, now, the fish season of the poor never, or rarely, knows an interruption. If fresh herrings are not in the market, there are sprats; and if not sprats, there are soles, or whitings, or mackarel, or plaice.[119]

The selection of available fish at Billingsgate Market "would enchant a Dutch painter" and reflected the diversity of marine

110

resources conveyed to London.[120] From the Atlantic waters of Cornwall to the North Sea harbors of Lincolnshire and Norfolk, such variety called for different prices and propelled new consumption habits. It was within this context that fried fish came to be Britain's most popular takeaway food in the twentieth century.[121] Already in 1851 Mayhew counted between 250 and 350 fried-fish sellers in the vicinity of Billingsgate, transforming portions of sole and plaice into "shapeless brown lumps of fish." Fish was chopped into indistinguishable chunks, battered with flour, and deep fried, which allowed it to be preserved and sold by street sellers distributing cheap food in poor neighborhoods.

Price guided the trade, and as a fish fryer affirmed to Mayhew regarding which species he favored: "I buy whatever's cheapest." This was "the overplus of a fishmonger's stock, of what he has not sold overnight, and does not care to offer for sale on the following morning." The leftovers were classified as "fryers" and sold to costermongers to upcycle. The practice secured a minimum revenue for the fishmonger and created an innovative product for an extensive urban market. Unlike dried or pressed fish, fried fish did not require long hours of preparation and cooking. The working class now had access to a cheap, ready-made, and protein-rich food. It was a perfect match between an uneven supply and a strong demand, and it soon defined entire city areas by the smell alone. The trade also gave rise to a new urban character, the fried-fish seller, who according to Mayhew typically had to live "in some out of the way alley, and not unfrequently in garrets; for among even the poorest class there are great objections to their being fellow-lodgers, on account of the odor from the frying."[122]

Business at Billingsgate boomed so quickly that in 1877, just twenty-five years after the opening of the modern market hall, it had to be replaced by a larger building. This new market, still standing today, was designed by Horace Jones (1819–1887), who succeeded Bunning as city architect. The open central hall is roofed by an ingenious scheme of top-lit iron roof trusses that contrast with the more discrete load-bearing brick walls of the surrounding market galleries. Billingsgate's distinctive working hours began at five o'clock in the morning, and by nine, the main business had been concluded. Several illustrations of the market interior show large oil lamps hovering over the stands, breaking through the obscurity of dawn by the Thames. But despite these facilities, critics were concerned with the hygienic condition of the building. Less than two decades after its opening, health associations complained that it was too small for its purpose and its sanitary conditions were "nothing more nor less than a disgrace":

> Instead of all of the internal fittings being made of glazed, varnished, non-porous, non-absorbent materials, most of them are substances highly absorbent, which breed every form of germ bacteria or microbe, specially active in starting and ulcerating the decomposition or putrefaction of any dead fresh fish which may be in the market.[123]

"Cheap Fish of St. Giles," London, 1877. Photo: John Thomson. From Adolphe Smith, *Street Life in London* (London: Sampson Low, Marston, Searle and Rivington, 1877).

Regardless of these concerns, and the harshness of the urban fish trade in the second half of the nineteenth century, fresh fish consumption continued to increase. Fish and chips, the popular pairing of fried fish with fried potatoes, was perfectly aligned with the indiscriminate catches of trawling, and in the early twentieth century the balance of supply and demand made it a long-term success, threatened only when frozen fish and new takeaway foods were introduced to the urban population. Particularly important, with regard to the ecological pressure placed on fish populations, was the adaptability of the technique of battering and frying to new varieties of fish, regardless of their palatable qualities. Because all fish can be fried, the popularity of fish and chips was great for the industry because it reduced offal and increased the profits of fishing, which was no longer required to target high-priced species.

With a new consumer culture came new ecological pressure. The growth of urban fish markets in the middle of the nineteenth century, facilitated by railway transport, was sustained by an increase in trawling and the development of new trawling technologies. The practice, which was not new, consists of towing a bag-shaped net along the seafloor. Trawling techniques varied widely depending on the type of vessel used, the nature of the seafloor, and the behavior of the targeted species. A leading harbor for trawling in the 1830s was Brixham on the English Channel in Devon, where wooden-hulled sailboats were the vessels of choice. The expansion of fish markets in the 1850s came with the exploitation of the rich fishing grounds of the North Sea, which moved the trawling centers from the Channel to older fishing communities such as Lowestoft and Yarmouth and to the newly developed harbors of Hull and Grimsby on the Humber River.

Several new technologies transformed the business. The first was the introduction of the mechanical capstan to wooden boats, which relieved fisherfolk from the hard effort of hauling a loaded trawl. As a result, boats could pull larger nets with smaller crews, operating faster over longer periods. Another innovation was steam power, which granted the boat independence in the face of variable winds. It also steadied the boat's movement, and this, combined with inventive devices to keep the trawl open and rolling across the seafloor, made underwater harvesting more efficient than in the age of sail trawling, when wind and currents needed to cooperate. Finally, new steel-hulled vessels incorporated the previous advantages to become powerful trawling machines. With navigational improvements and operational advantages over older boats, the new vessels could take larger hauls over longer periods far from their landing harbor. As Callum Roberts pointed out, whereas the early "sailing trawlers were 20 to 30 tons burden," those of the 1870s "were larger—70 or 80 tons—longer and leaner and carried a much greater spread of canvas, lending speed and power."[124] Grimsby and Hull were the harbors that best represented this "trawling revolution," a change made possible by significant growth in the urban demand for fish.

Kéroman, the fishing harbor of Lorient, is a telling demonstration of how the modern harbor created a fundamental shift in fishing activity, changing its impact on marine ecosystems. Until Kéroman was built from scratch between 1919 and 1927, Lorient was not an important fishing center—France's leading harbor was in the north at Boulogne-sur-Mer.[125] Whereas most fishing harbors are components of larger infrastructures with multiple functions, Kéroman's developers built it with the main purpose of scaling up the fishing industry. This specificity and Kéroman's location within the complex network of Brittany's sardine fisheries renders visible how, motivated by the ecological limitations of the sardine business, investors moved their capital from factories to the harbor. As in England, the railway was a factor in this change. An efficient harbor needed rail infrastructure before it could become a hub that channeled abundant supplies of fish from the coast toward a growing urban consumer market. While the harbor and the factory were complementary, they had different ecological

overleaf Billingsgate jetty looking toward Tower Bridge, London, ca. 1930. Photo: George Davison Reid. Courtesy Museum of London/George Davison Reid Collection.

footprints and involved different actors. Because of the astronomical sums of investment capital required, harbors were run by bankers and developers with state support, not by fisherfolk. In sum, the modern harbor acted as a threshold that separated artisanal fisheries from industrial ones. As an architectural device, it facilitated an exponential increase in fish landings and detached extraction from consumption. Kéroman is a good illustration of this industrial turn.[126]

Seen from above, Kéroman has five key components: a basin, coal docks, an ice freezer, a processing hall, and a railway terminal. Together these elements suggest how fundamental consistency and time management are in maintaining a stable flow of fish from sea to city. Coal docks fueled steamships that could trawl continuously instead of depending on the moods of the wind. The freezer provided ice to slow down the decomposition of the fish, which allowed enough time to complete the operations necessary to deliver it to market while still fresh. The basin and the railway worked together to move product in and out of the harbor at a constant pace. The basin maintained a navigable water level even at low tide so boats could land fish around the clock. The regularity of this activity was mirrored by the regular rhythm of loaded freight trains leaving the terminal. And, finally, the hall was where fish were unloaded and processed, a pivotal point of transfer between systems. Kéroman was purpose-built to host these sequences of synchronous operations. With a dedicated area for each function, the site had all the ingredients of a technological landscape. Engineer Henri Verrière based the underlying logic of the design on a five-to-one ratio derived from the conversion rate of coal to fish: steam trawlers had to burn five kilograms of coal to obtain a kilogram of fish.[127] Kéroman's south mole, a mechanical apparatus that efficiently unloaded coal from cargo ships docked on the outer side of the harbor and then loaded it onto trawlers in the inner basin, must have excited modern architects with its hoists, electric cranes, moving gates, railways, and a fine coat of coal dust.

Unlike the canneries in Brittany that evolved from sardine-pressing workshops and the anarchic proliferation of fish fryers around Billingsgate, each the consequence of consecutive developments in the fishing industry, Kéroman's design was based on careful planning. Not only did its founders learn from nineteenth-century harbors like Grimsby and the more recent installations at Cuxhaven in Germany and IJmuiden in the Netherlands, but they

also made a careful study of context. The geographic location proposed was close to the overlapping habitats of many species of fish at the edge of the continental shelf, and this advantageous position helped local backers to maneuver larger national investors into securing the site. Many contemporary chroniclers emphasized the uniqueness of the opportunity to build a harbor from scratch, near Lorient but not encumbered by preexisting urban conditions, close to fishing grounds and experienced fisherfolk but not an outgrowth of an established fishing community.

At its launch Kéroman was an ideal ex nihilo architectural machine for the processing of fish. Yet coal steamers would soon be replaced by diesel engines, making the state-of-the-art coal docks obsolete. The fishing grounds that nurtured Lorient's trawling fleet began to show signs of exhaustion, pushing the trawlers farther into the Atlantic and North Sea, which consumed more fuel.[128] Details of the connections between these occurrences have yet to be established, but a chart of landings at Kéroman published in 1944 shows a consistent variation in the species being captured and in the relative importance of each type of fish.[129]

Kéroman fishing harbor, Lorient, 1927. Courtesy Archives Municipales de Lorient, 5Fi3313.

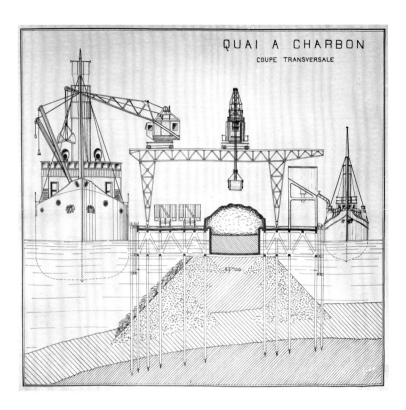

QUAI A CHARBON
COUPE TRANSVERSALE

Hake dominated the early 1920s, representing up to 45 percent
of total catches, but by the beginning of the 1930s, its share had
dropped to 17.8 percent and rays took the lead.[130] The changes in
targeted species correspond to the exploration of different fishing
grounds, and this implied technical and economic adjustments.
But even as ecosystems and extractive dynamics were transformed,
Kéroman stood still, creating an unbalance that highlights a dis-
connect between ecological cycles and the time of architecture.
Once the construction was finally in use, the ecosystems that justified it had already moved on.

As with any massive infrastructure project, the development of
Lorient's fishing harbor depended on the willingness and engagement of many actors. One of the notable locals who had their contributions memorialized with a street name is the industrialist
Émile Marcesche (1868–1939). His story helps to illustrate the complex dynamics at work in the construction of a modern harbor.
Lorient was surrounded by fishing communities, but fisheries were
not the focus of its economy until Kéroman was built. Founded
in the seventeenth century as a base for the French East India

118

Company, Lorient became an important naval harbor in the eighteenth century.[131] Marcesche, a newcomer to the city, married into one of Lorient's wealthy families and went into partnership with a local coal trader in 1898.[132] An astute businessman, he attracted local investment to support a profitable import-export business trading timber mine poles for Welsh coal that benefited from the Paris-Orléans rail connection. The operation required a significant fleet of cargo ships and expertise in long-distance navigation. By 1904 Marcesche had diversified his business interests to include a trawling company that operated in tandem with the coal business.[133] He recognized the strategic advantages of Lorient's location for exploiting the rich fishing grounds off the coast of Brittany, describing them as "an immense area over which steamers never cast their dragnets without profit."[134] And as a coal dealer, he could leverage the complementary opportunities the fishing industry presented to provide fuel to the steamers and increase his profits.

One of his associates in the harbor venture was the banker Auguste Ouizille (1853–1943), who with his brother Georges (1861–1945) had important investments in the Delory and Colin canning empire.[135] Frédéric Delory (1840–1892), the founder of the eponymous company of sardine canneries, had recently passed away while in office as mayor of Lorient. In the 1880s, Delory led the industry in his company's twofold response to Brittany's sardine crises. First, he diversified by expanding into canning vegetables. Second, he sought alternative fishing grounds and relocated production units to Portugal and Algeria. Following the regime shift that collapsed Brittany's sardine populations in 1880 and 1881, Delory secured partners in Setúbal, south of Lisbon, and together they set up an industrial operation that gave the Portuguese canning industry a significant boost.[136] Over time, they expanded into the Algarve region. At its peak, Delory's empire counted six canneries in Portugal, two each in Morocco and Algeria, and another six in France. The wages paid to Portuguese workers were lower than in France, which was a particular advantage since high payrolls in France had exacerbated the sardine crises there. But ecological factors were also important in establishing production units in Portugal. Not only did the regime shifts of the 1880s hit Portuguese sardine populations less severely than Brittany's, but the strong upwelling off the Portuguese coast provides sardines with abundant food and gives them a unique taste. By operating internationally, Delory's canneries are an early example of the gap

Coal trawler in the Kéroman fishing harbor, 1935. Courtesy Archives Municipales de Lorient, 7Fi685.

overleaf Henri Verrière, plan of the initial project for the Kéroman fishing harbor, Lorient, 1924. Courtesy Archives Municipales de Lorient, 9Fi2817.

Légende

Voies ferrées (le rayon minimum est de 120ᵐ)
Travaux projetés
Terrains à acquérir
Rues principales accessibles à tous véhicules
Quais accessibles aux chariots électriques seulement
Profondeur à obtenir au dessous des plus basses mers

Échelle : 1/2000

Agrandissement éventuel.

Bassin éventuel

Château de Kéronny

Limite

Rivière Le Ter

LORIENT

Nord

Agrandissement éventuel

Port de pêche projeté

between exploited ecosystems and the source of the capital that supports that exploitation, a pattern that would soon be repeated as new harbors were built to support trawling fisheries.

It is often claimed, without proof, that the construction of Kéroman ex nihilo was made easier because there was no preexisting fishery in the area, meaning that there were no local fisherfolk to fight modernization. But, in fact, the new harbor depended on preexisting elements such as the railroad that came to Lorient in 1862 and the coastal sardine fishery. The development of the steam-powered trawling industry was aligned with coal interests, and Marcesche, with his coal business and his fleet of steamers, was in an ideal position to benefit. His association with the Ouizilles meant that the capital that financed the harbor came from the sardine canning industry. Having weathered the seasonal ups and downs of sardine catches, the Ouizilles and others were anxious to find an alternative fishing model that would keep France's fishing industry competitive. Although Breton fisherfolk did not initiate the shift from artisanal to industrial fisheries, their presence and the capital that industrialists accumulated from their labor were crucial in building an industry that exploited fishing knowledge through new means and technologies.

Although the trawling business was funded by private developers, it required heavy infrastructure that only the state could provide. As prominent industrialists, Marcesche and his associates were well-placed to lobby for such investment, although it took time. While Marcesche started to develop a trawling fleet in 1904, it was not until 1917, during the Great War, that the French government appointed the engineer Henri Verrière to direct the construction of the new fishing harbor. Deliveries of frozen fish to the war front had become essential in keeping the French army fed.

As Verrière understood, speed was the goal. He designed the harbor to expedite products by carefully designing each one of its processes—docking, unloading cargo, trading, and loading carriages—in efficient coordination. Verrière envisioned the harbor's components as forming "a fish market located at a train station."[137] Thus the natural susceptibility of fish to spoilage met its match in the modern love for speed and timed operations.

Prior to the development at Lorient, the neighboring city of La Rochelle, whose old port had long fostered a significant fish trade, built a new harbor a few kilometers west of the city.[138] The idea was to boost the city's chemical industry by combining a harbor with

Coal docks for the Kéroman fishing harbor, Lorient.
Courtesy Archives Municipales de Lorient, 7Fi698.

an industrial park built over an "empty" marsh area.[139] Originally
designed by engineer Anatole Bouquet de La Grye (1827–1909) in
1877 and christened La Pallice, the harbor featured a large water
basin served by railroads and provided ample surface area for the
construction of new warehouses and buildings to support a flour-
ishing industry. A survey drawing shows the inherent logic of the
operation: red lines over the existing terrain mark where the basin
would be dug, and the hatched area north of it indicates where the
excavated soil should be deposited to form an embankment. This
massive project transformed an intertidal marsh into industrial
lots served by the railway, shaped a new coastline, and connected
the dynamics of the sea with the hinterland markets. Yet La Pallice
was not for the fishing industry, which continued to operate from
the wharves of La Rochelle's old port.[140]

It was not long after the completion of La Pallice, at a time
when steam trawlers had begun to operate in the area, that the
marshes surrounding Lorient were considered for a similar investment focused on the fishing industry. An oft-quoted passage by
Verrière describes a stroll the engineer used to take near Lorient
in 1904, during which he would "pause happily at a rocky point"
that "provided direct access to deep water." He imagined that with
"the construction of a pier" it would be just right for mooring "the
steam trawlers that, at low tide, were beached in the Lorient harbor."[141] Verrière's memoir recalls a passage on Goethe's *Faust* in
Marshall Berman's study of modernity, which points out how, with
the support of Mephistopheles, Faust changes from a "dreamer"

and a "lover" into a "developer."[142] As Berman tells it, exhausted from many extraordinary adventures, Faust and Mephistopheles "find themselves alone on a jagged mountain peak staring blankly into cloudy space, going nowhere." Then "suddenly, Faust springs up enraged: Why should men let things go on being the same way they have always been? Isn't it about time for mankind to assert itself against nature's tyrannical arrogance, to confront natural forces in the name of 'the free spirit that protects all rights'?"[143] At this point Faust unfolds plans for dams, canals, harbors, and other infrastructure, and then organizes titanic workforces to transform the land and sea into productive industries and cities.[144] This could have been Lorient's developers as they built Kéroman.

For decades, Marcesche was the harbor's biggest advocate, and from within a complex web of commercial alliances between various partners and competitors, his trawling company dominated Lorient's fishing business. In 1906, Marcesche's companies operated six steam trawlers; in 1908, the number tripled to nineteen; and in 1926, his most profitable year, there were fifty. He became famous for promoting the "Quinzaine du Poisson," a two-week festival that celebrated eating fish and encouraged urban consumers to develop new food habits. And he served as the head of the Bourse du Commerce, an entrepreneurial association that stimulated public investment in the fishing harbor. It was in this role that, in 1926, he set up a fish-reduction plant that would later become an operational and financial disaster.[145] He subsequently lost his concession to operate the harbor, and along with it the power to control conditions that affected his multiple interests in the fishing business. The fact that he let the harbor slip from his grasp has intrigued historians, and it was certainly the outcome of rivalries and power struggles behind the scenes. In the harbor, decisions were made independently of the ecosystems being exploited.

Luchino Visconti's 1948 film *La terra trema* portrays the challenges faced by traditional Sicilian fisherfolk and the risks and responsibilities of boat ownership.[146] Visconti's neorealist vision focuses on the dynamic between the fisher as an individual—subject to the pressures of wholesalers and capital—and ownership of the means of production, whether by an individual or a cooperative. It wasn't so different in France a generation earlier: in 1917, the head of the French fisheries office commented that "the fisherfolk had henceforth to say goodbye to the small family home and to fishing with his own children, goodbye to the complete freedom he enjoyed on his own boat; he has been seized by the industrial machine and is now a simple cog within it."[147] While boat ownership certainly creates a social divide within the fishing industry, from an architectural standpoint, ownership of infrastructure is even more critical. As fishing harbors were mechanized and the pace of the fish trade accelerated, the value of the extracted resource went down as the volume of consumption went up. The growing gap between the price and the size of the market emphasized social divisions within the fish trade and, even more significantly, by putting emphasis on the monetary value of total fish landings, ignored the ecological value of a fish as an animal in an ecosystem. As food markets expanded, entrepreneurs replaced fisherfolk in the management of fish ecosystems. The harbor changed the players, from the boat owner to the industrialist, and in doing so changed the relationship between land and sea and flipped the social status of fish as a natural resource.

Construction of Kéroman cold storage, Lorient, 1927.
Courtesy Archives Municipales de Lorient, 5Fi3318.

4   *The Salt and the Freezer*

Imagine a fish's-eye view of architecture. It differs from the classical bird's-eye view that was the prelude to modern urban planning.[1] The bird's-eye view allies itself to Le Corbusier's (1887–1965) ghost, his fascination with machines and airplanes, which would change the mechanics of architecture and the standpoint from which urban transformation was observed. That, at least, was historian John Summerson's (1904–1992) opinion when, in 1945, he asked the reader of his *Georgian London* to imagine himself "suspended a mile above [the city], staying up there for a period of time proportional to two centuries, with the years speeding past at one a second." As he argued, "The spectacle below you proceeds like those nature films which accelerate into immodest realism the slow drama of plant life. The life of a city, condensed so, would be as dramatic. It would give the same startling impression of automatic movement, of mindless growth."[2] Photography took the fish's eyeball as a reference for the wide-angle lenses capable of distorting perspective and reality to depict enclosed spaces. Yet the fish's standpoint is underwater, and once out of this environment, it soon dies, often losing its head to the fisher's knife.

From the perspective of fish, architecture is a passage. Appearing on its path from animal to commodity, buildings are interfaces, facilitators inscribed within a sequence of logistical movements leading from their harvesting to their arrival in distant places to be turned into food. In this process, no building typology has had a larger impact on ecosystems than freezers. A freezer is a technical device for removing heat and is not necessarily a building, but when such devices were incorporated as key elements in the landscape of fishing, a wealth of buildings were designed to accommodate refrigeration technology. These producers of cold started to populate harbors, railway junctions, and cities on the eve of the twentieth century, and they were like nothing that fishes had seen before. At the same time, in a parallel development, sailing boats started to be replaced by steamers in the waters of fishing harbors, and chimneys took the place of masts. Mammoth blocks consisting of multiple levels, freezers introduced a new sense of verticality into port panoramas, a verticality that was once the preserve of masts. Often painted white, freezers contrasted with the smoke of coal—of which they were the logical outcome—their new shapes in the harbor dialoguing with the clouds emanating from trawlers in the next stage of what was termed the "modernization" of fisheries.

Countering bacterial putrefaction and slowing the decay of fish, as indicated by organoleptic qualities such as appearance, odor, and taste, cold aims to balance both market fluctuations and the seasonal and circumstantial ups and downs of natural resource extraction. While homogenizing the cycles of fisheries and standardizing the capacity of producers to feed consumers, refrigeration contributed to the safety of capital investments and helped regulate the natural processes of flux within ecosystems that affected patterns of consumption. It affected the culture of transporting fresh fish and generated new habits of consumption geared to frozen fish, even giving rise to novel products like fish sticks.[3] Hence, by enabling increased consumption and exerting greater pressure on fish populations, refrigeration and freezers brought incomparable transformations to both the landscapes of fishing architecture and the ecosystems that they prey upon. They became a key in the "complex mingling" of "first" and "second nature," between the "prehuman nature" and the "artificial nature that people erect atop first nature," to use William Cronon's terms to understand the connection between the urbanization of modern cities and the vast ecological territories they draw upon.[4] From an economic perspective, the cold storage was an instrument for food speculation, creating the conditions for a futures market of perishable commodities. More than mere technical apparatus and epitomizing the human drive to tame nature, freezers went beyond technological determinism and became part of major cultural movements.

A freezer can be either a refrigerated warehouse, a cold storage, or a factory to process frozen goods. Its origin as a building typology is only marginally connected to fish, and its history draws on manifold usages and technological developments adopted from various sources. As connectors linking different ecosystems, freezers are part of technical infrastructures—nodes within what has become known as the cold chain—and rely on preexisting infrastructural networks. A railway, an electricity supply system, and a harbor are foundational elements of primary infrastructure systems, on which depend secondary systems such as the cold chain.[5] The freezer is a hub of the cold chain, the latter being a sequence of different communicating vessels connecting the fisher who extracts the fish from the sea and the consumer who brings it to the kitchen. The domestic refrigerator, the fishmonger's stall, the distributor's isothermal vans, the refrigerated railway wagon, the warehouse, the freezing factory, and the boat that transports the fish are all channels and nodes, systems of communicating threads whose relative coherence makes cold chains stronger—or weaker. Their goal is the rationalization of transport and storage, a key element in strategies designed to master natural cycles and level out fluctuations between supply and demand.

Susan Freidberg has demonstrated how the idea of fresh food is a cultural construct.[6] The scientific acknowledgment of vitamins' role in human health was a key factor in boosting refrigeration, while the continuous supply of ice in urban areas accompanied a growing concern about dietary habits: eating better was the best remedy to counter the rapid spread of uncontrollable diseases like tuberculosis. If food's nutritional qualities depended on its freshness, there was a general desire to safeguard its healthy properties and preserve it more effectively. By the late nineteenth century, the icebox, the predecessor of contemporary refrigerators, emerged as the urban manifestation of the novel fresh food culture. A cold industry developed that aimed to cater to these multiple burgeoning markets.[7] Historians agree that breweries and the perfecting of beer-making boosted the consumption of both "natural ice," harvested from lakes and rivers, and "artificial ice," produced by refrigeration machines. After beer, meat contributed to the development of cold technology. Meat's taste and qualities improve from the slight decay that takes place after animals have been slaughtered, and it was the meat industry, which pushed the development of new technologies, that benefited from the possibility of connecting distant areas of production. As boats were sufficiently large and stable to accommodate the complex machinery required to produce cold, marine traffic made it possible for distant places like the Argentinian prairies to be connected with the British market. In concert with the meat market, fruit and vegetables played a significant role in precipitating the adaptation of railway wagons to carry refrigerated goods.[8] In this context of developing markets and technologies, fisheries contributed an ever-growing demand for ice, which was needed to facilitate the transport and distribution of fresh fish, especially in the developing trawling sector, which found, in ice, an ally to support larger catches farther out to sea.

Ice falls from the sky, and with negative temperatures water freezes. Hence, there were locations where for a long time natural ice was used in fisheries, and important structures for harvesting

and preserving ice were developed.[9] While in many places frozen water could be obtained locally, in the nineteenth century an
important international ice market developed, with English fisheries acquiring hundreds of tons of Norwegian ice. In the United
States, where the icebox become a domestic standard, the trade
expanded into a very profitable business, and a new architectural
typology emerged: the ice elevator. These elevators were practical structures, many stories high and built in wood, with efficient
interior partitions and exterior double walls filled with sawdust or
other insulation materials: they were used, often with the help of
mechanical conveyor belts, to convey heavy blocks of frozen water
to the top of these monumental constructions. By the end of the
nineteenth century, there were plenty of these wood constructions
on the margins of Massachusetts lakes, catering to the desire for
freshness not only of Bostonians but also of the inhabitants of the
fishing villages dispersed along the Atlantic shore. Their existence

Icehouse burning, April 6,
1930. Courtesy Robbins
Library, Arlington
Historical Photograph
Collection.

Quincy Market Cold Storage, the "Boyle" refrigerating machine, made by Pennsylvania Iron Works, Boston, 1895, in *Ice and Refrigeration* 9, no. 6 (December 1895): 383.

was short-lived: soon the "natural ice" harvested in lakes and rivers was to be replaced by "artificial ice" produced by the quickly evolving refrigeration machinery. Artificial ice was cleaner than harvested ice, its production was less dependent on weather conditions or seasonal variations, and its continuous output obviated the need for gigantic storage facilities.

These icehouses burned easily; fires were such a common sight in New England icehouses that insurance companies were reluctant to underwrite them. Built in wood and filled with sawdust, they were highly flammable and quickly consumed by fire. Newspapers pointed to two triggers for this: one, deemed more frequent, was spontaneous combustion—"when the warm weather comes on the moisture from the ice working out through the dry packing causes combustion and a fire follows"; the other was the "carelessness of the men employed," who "enter the houses smoking their pipes, and, feeling sleepy, spill the ashes into the

packing and a fire follows."[10] In May 1894, in Arlington near Boston, a 213-meter-long icehouse on the shore of Spy Pond burned to the ground. A newspaper reported the sublime event, as the light of the fire reflected off the calm surface of the lake water: "When the four immense houses were in flames the sight was a most magnificent one. Innumerable boats filled with occupants were out on the lake viewing the fire, everyone standing out in bold relief against the darkness of the night."[11]

A new typology soon followed that was less prone to burn: the cold storage. Warehouses were adapted for refrigeration either with the use of ice or through novel forms of machinery. Within these buildings erected for cold, there flourished a new inner world of icy and frozen spaces. From an architectural perspective, these were the ultimate interior space, enclosed rooms deprived of natural light, with low ceilings and an endless network of pipes and frosted coils, that were used to stockpile fish-like white objects stored in a dim environment. Emphasizing this uncanny interior, the cooling coils led to the construction of gigantic machinery rooms: larger-than-human dark devices with lubricated pistons, valves, gear wheels, engines, and intricate arrangements of forms occupied halls flooded with natural light. It was the machinery, not the building, that created cold. Refrigerators weighed hundreds of tons, with compressors, condensers, generators, heat exchangers, receivers, and absorbers, steamed by boilers which in turn fed on coal that spread a thin coat of dust over a predominantly white landscape. These machines, reaching a height of several stories, were part of the novel imaginary of modern times, maneuvered either by clever engineers or by Charlie Chaplin's docile wrench. As Oscar Anderson, the pioneer of refrigeration history, remarked, "The ice factory captured the imagination of all who saw it operate. Coal was brought in at one end of the plant; blocks of crystal-clear ice were removed at the other."[12]

Architects were soon involved in the making of this chill new world. One of the companies producing these heavy refrigerating machines was the Hercules Iron company of Aurora, Illinois, which in 1893 supplied the systems for the Cold Storage Building at the World's Columbian Exposition in Chicago. Designed by Franklin Burnham (1853–1909) in association with his namesake, the master Chicago architect Daniel Burnham (1846–1911), the exposition's cold storage was an elegant monument, an opaque plastered structure with a peripheral gallery on its top floor and a 67-meter-high central tower. Located by the southwestern entrance to the fair on 67th Street, behind the railway tracks leading to the exposition's Terminal Station, the attraction had a large and glamorous ice-skating rink on its top floor:[13] it was equipped to store and supply fresh food to the fair, and its interior "displayed the various methods of artificial freezing."[14] The floor plan, with a 40- by 78-meter footprint, had a tripartite structure: its north aisle provided cold storage that was rented out to exhibitors (generating 40 percent of the operation's income by storing goods ranging from chocolate sweets to watermelons along with "some" fresh salmon); the south aisle hosted the ice-freezing machine (100 tons daily production, sold at the expensive rate of $4 per ton, and generating 30 percent of the operation's income); and the central area hosted the heavy and expensive machinery.[15] The central tower was the system chimney, and like the pavilion's metal structure, it was covered with wood and plaster to give the construction some architectural glamour. On July 10 a tragedy happened.[16] The tower caught fire and reduced the pristine white building to ashes, costing the lives of twelve unfortunate firemen and three civilians. Although the causes of the fire were never established, there were two common explanations: one was the ephemeral timber and board architecture, which was independent of the metal structure and allowed the airflow to quickly spread the fire throughout the building, catching the fireman by surprise; the other was the implausible "deafening explosion of the ammonia pipes" used in the refrigeration system that many within the horrified crowd guarantee they heard.[17] Regardless of the causes, cold storage facilities and their buildings were on the front pages of daily newspapers; their architecture was noticed and the advantages they brought were at the risky vanguard of the "modern world."

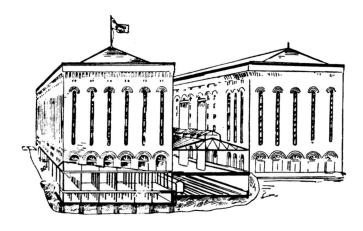

Assessing the role played by architecture in developing systems for environmental and economic regulation, the historian Michael Osman singles out two examples of early cold-storage systems built in the United Sates and how architects dealt with their designs: the Chicago Cold Storage Exchange, designed by the partnership of Dankmar Adler (1844–1900) and Louis Sullivan (1856–1924), and the Boston Quincy Market Cold Storage Company, designed by William Gibbons Preston (1842–1910).[18] Osman points to Adler and Sullivan's attempts to develop a language for a novel public monument, designed to incorporate mixed commercial functions, office space, and welcoming arcades alongside the refrigerated storage directly served by railway spur lines. Echoing

Adler & Sullivan, Cold Storage Exchange, Chicago, 1890, in *Chicago Tribune*, November 14, 1890.

Franklin Burnham and Daniel Burnham, cold storage building at the
World's Columbian Exposition, Chicago, 1893, in *The Book of the Fair*, 1893.

the architects' contribution to the novel typology of the steel-structure skyscraper, the generous public areas and the care taken in the building design contrasted with the quick economic debacle of the enterprise and the later demolition of the complex. The project started in 1890 based on the application of the technology for compressing and circulating ammonia, creating rooms with regulated temperature by means of thermostatic mechanisms. The venture aimed at nothing less than being the "biggest in the world": two ten-story-high blocks of refrigerated warehouses—116 meters long and 20 meters wide—located upon a railway spur with direct access to the Chicago River. Its size was matched only by its transitoriness: in 1895 the company went bankrupt, and in 1902 demolition started. The few surviving images of the massive monument remained marginal within Adler and Sullivan's body of work and the canon of modern architecture.[19]

Unlike the failures of Chicago's famous experiments, a precursor in Boston with a different architectural pedigree had been a success.[20] The Quincy Market Cold Storage, a six-story block with a 50- by-25-meter footprint, was smaller than its Chicago peers but large nonetheless. The project was launched in 1882, at the junction of Richmond Street and Commercial Street, between the company's quarter in Eastern Avenue Wharf and the central Boston Quincy Market—from which the cold storage company took its name—and a few blocks north of the active T Wharf, where the important local fisheries were based. Preston designed an elegant brick building, with five arches for doors and windows set into a composition dominated by pilasters vacillating between monumental orders. As the magazine *American Architect* described the project, "a very large quantity of one-inch-thick hair-felt is used in the construction and insulation of the interior of the building. All the walls are built hollow, and a portion with double air-spaces; windows are in triplicate, doors felted, and special precautions taken with rendering the roof non-conducting."[21] The storage's refrigeration system relied on ice alone. Hence, the building section is revealing, a sturdy steel structure with raised floors and cross bracing was topped with a double-height upper level where the cooling ice was stored, the latter occupying approximately half of the storage volume.[22] This system required careful drainage arrangements to collect and evacuate the melted water without compromising the stored goods. The operation was so successful that in 1890 the company initiated an ambitious expansion plan,

William Gibbons Preston, Quincy Market Cold Storage Company, Commercial Street, 1882.
Photo: Richard Merrill. Courtesy Boston Public Library, Arts Department.

doubling the available cold storage area and replacing the ice-refrigeration system with new machinery that started operation in spring 1892. Adjacent to the building, a new boiler house and a larger engine house hosted the complex refrigerating machinery. The compound kept growing, and soon after a new building was erected beside it to extend the floor plan; the original structure was reinforced to accommodate a new floor on top of it, while the now-obsolete storage for cooling ice was subdivided.

Eggs, butter, vegetables, and meat made up the major share of Boston's cold storage business. Yet fish was among the frozen products that were stored using the "direct expansion system." To be preserved in cold storage, fish needed to be frozen, and this required specific procedures: "The direct expansion system is used in the rooms in which fish are frozen, the fish while freezing being hung on racks, and in order that the racks may be used for other lots of fish without too great delay it is necessary that the fish be frozen hard in from twenty-four to thirty-six hours."[23] Over the following years new premises were acquired and rented on the Eastern Wharf and Clinton Street. When in 1895 *Ice and Refrigeration* magazine dedicated a long piece to "Boston's Cold Corner," a pipeline infrastructure was being installed under Commercial Street alongside other urban infrastructures such as gas, electricity, and telephone. The connection allowed refrigerated brine to be exchanged between buildings, a network that prompted the concentration of machinery and increased the power of the company. In the spring of 1906 it opened a new building, which was also designed by Preston, this time a ten-story block of 20 by 50 meters, the insulation consisting of "galvanized iron cans filled with rock wool" combined inside a wall made of "glazed terra cotta blocks."[24] The new structure was enlarged again in 1908, making it possible "to devote the Richmond St house, with a capacity of 1,000,000 cubic feet, exclusively to the freezing and storage of fish";[25] and in 1916 the company acquired T Wharf, the former fish pier, to erect another new colossus, combining nine-story-high office spaces with ten-story-high cold storage.[26] This continuous urban expansion did not target fish, which in 1915 only accounted for about 10 percent of the Quincy Market storage space.[27]

Preston's pragmatic designs for cold storage spaces were a response to the quick physical transformation of the urban food landscape. It was not only food that was being transformed

143

through refrigeration and new concepts of freshness: its infrastructure required new design strategies, incorporating innovative technologies and responding to a different urban culture. The removal of the former T Wharf fishing pier and the construction of a new, "modern" facility to replace it in the South Boston harbor expansion is a case in point. T Wharf, near Quincy Market's cooling empire, had been a reference point in the urban landscape for decades: "Very few visitors to Boston ever left the city satisfied unless they had made a visit to this fish pier to see the fast schooners, clever fish handlers, splitters, etc., and the various kinds of fresh fish to be seen there."[28] It was said to be served by 325 vessels, employ 6,500 men, and trade 56,700 tons of fish annually. Always emphatic, local newspapers would claim it was "the second largest fishing port in the entire world," while reluctantly acknowledging that Grimsby, in England, was the largest.[29] The premises had functioned for the last thirty years on a preexisting wooden pier, where a long, three-story-high row of stalls served the merchants in the cramped docks. The move was part of a larger harbor expansion toward the south, the new docks offering relief in an area where urban pressure was growing, a pressure confirmed by the swift sale of the premises to the cold storage company.

The move from T Wharf to the developing South Boston harbor was agreed in 1910.[30] It was decided that a new dock, about 90 meters wide and 370 meters long, would be built, with landfill between retaining stone walls granting 7 meters of water at low tide. While the pier facilitated navigation and offered better landing options, three distinct buildings hosted the main functions of a modern fishing harbor: a stock exchange, merchants stalls, and a freezing facility.

On top of the pier was a prominent detached construction facing out over the water: the Fish Stock Exchange. It was marked by a monumental archway that gave access to the trading floor, where business was conducted in relative proximity to, but still independent from, the boats and the docks. This hygienic detachment increased the gap between commodity and animal, and the independent building contributed, through its architectural form, to fish being conceived of as an abstraction. The design was entrusted to the Boston architect Henry Francis Keyes (1879–1933), who chose a classical language with a rusticated ground floor, a

144

pediment adorned with fish motifs, and a strong entablature, which gave the construction a powerful, solemn presence within the functional architecture of the harbor facilities.[31]

The classical language of the stock exchange was extended in the two three-story-high rows (18 meters wide and 230 meters long) of stalls designed to cater to forty-four fish merchants. Newspaper descriptions gave a vivid idea of their architecture: "It looks solid to the eye, but inside its brick wells, under its concrete flooring, deep in its granite bowels, it is tunneled and tubed with drainage systems, plumbing and purgative devices."[32] Their organization, including "a 'cooler' or ice box" was simple and clear: "Each of the stores has its office, downstairs, where the cashier reigns supreme. There is as well an office upstairs where the heads of the firm hold forth. The store proper is lined with tile, with a concrete floor. There is a drain in the center of each and a hydrant through which salt water is drawn to wash the place out each

Boston Fish Pier, 1930. Photo: Fairchild Aerial Survey. Courtesy University of Massachusetts Boston, Joseph P. Healey Library.

night."[33] The sophisticated appearance of the buildings surprised contemporaries, who nevertheless agreed with the architects' idea:

> Merely because an establishment is primarily for use is no reason for creating it ugly as well, and, furthermore, that to achieve ornamental effects does not necessarily incur the expenditure of large sums on ornate gingerbready. The dignity of these buildings is in their lines and in their proportions, yet dignity is theirs, in such measure, that, seen in groupings, they seem more like exposition buildings for a permanent world's fair than humdrum marts of trade.[34]

The third and most monumental of the fish pier constructions was the freezer. Unlike the exchange hall and the fishmonger stalls, this larger building was devoid of architectural moldings. Its sheer volume and the exposure of the concrete structural grid would have enchanted modern architects and historians: to use Reyner Banham's reference to American industrial buildings, the freezer belonged to a concrete Atlantis.[35] The massive block had a U-shape form, enclosing the two rows of fish stalls and creating a monumental front to the desolate terrain, through which railroad spurs connected the complex with the logistical networks.[36] Its footprint was already marked in the 1910 plans of Boston, with a 68-meter-long, eight-story-high façade, with a double-story portico granting access to the inner courtyard of the fish pier. The plans make explicit that the spur tracks directly connect the fish pier with the railway system. A photograph taken when the pier started its operations shows the main façade of the massive structure under construction, the slabs' transparency contrasting with the opaque surfaces of the completed volumes and the formal effects of the fishmongers' stalls. The image is populated with carts pulled by dozens of horses, a form of animal traction that recalls the mechanical transformations that were taking place while the freezer was being built.

On Saturday, March 28, 1914, the "T Wharf flag was hauled down while a band played, and the T Wharfers took off their hats," marching in a procession headed by a band to the new Boston Fish Pier. The organizers "had 'Billy' Curran, one of the buyers, all decked out in a brand-new suit of oilskins and astride a horse to lead the procession. Every team was decked out with flags and almost everyone carried one or more big placards."[37] By Monday morning, twenty-two fish schooners were landing 680 tons of fish, most of it haddock, cod, pollock, hake, cusk, and halibut, while a shipment of haddock arrived by train from Provincetown. A beam trawler had brought in its nets the first five fresh mackerel, whose presence "in New England waters so early is considered a good sign."[38]

Fish exchange, Boston Fish Pier, ca. 1915. Photo: Ed Fitzgerald.

While the fish exchange's new facilities smoothed the processes of trade, the freezer provided the technical means to quickly transform the perishable goods into durable assets. Descriptions of the building stressed that "all walls, floors and partitions are of solid construction, eliminating every 'dead air space' in order to leave no breeding place for germs." The building, "designed [to be] as nearly automatic as possible so that in the event of labor troubles operations could be continued indefinitely with only a few men," combined cold storage capacity, a freezing plant, and an ice factory. In the cold storages fish was "kept many degrees below zero," after being subjected to "sharp freezing," a process in which "each individual fish is [encased] in an air-tight jacket which prevents the cold dry atmosphere in the storage chambers from evaporating the moisture in the fish, and also prevents germ infection." The ice factory had an "immense room" with an 18-meter-high ceiling to store the artificial ice, which was also to be automatically delivered to the boats and stalls: "From the bottom of the ice crusher the fine ice is shot by gravity through large tubes into hopper steel cars on an electric railway and is delivered to the various dealers through tubes into convenient ice boxes on the lower floors."[39]

While the cold storage business prospered in Boston, catering to eggs, butter, dairy, and other perishable products, the fish pier's colossal freezer was entirely devoted to supporting fisheries and the fish trade. Its purpose was not focused on local business nor simply oriented to the pier catches but aimed to function as a regional hub, receiving and distributing deliveries from Massachusetts as well as from Nova Scotia and Newfoundland. The chronology of this building type is one of its most interesting aspects. Despite being contemporary to other cold storage facilities and freezing infrastructure, it preceded the routine consumption of frozen fish products. The use of crushed ice mixed in successive layers of ice and fish carefully laid in crates, preventing fish quick putrefaction and increasing its commercial longevity, started to be common in the mid-nineteenth century, when

148

shipping fish by rail became standard practice.[40] It was quite a different matter to freeze the fish, either entirely or dressed, a process that required extremely low temperatures and quick freezing to prevent the fish from losing too many of its qualities. At the time, the relatively rudimentary systems deployed were experimented with in various locations. Different fish species reacted in a variety of ways to the freezing operation, with the Great Lakes herring and the Pacific salmon fisheries taking the lead in the business. In the Atlantic area, an important aspect of this initial development was the preservation of fish for bait. Between 1890 and 1893 freezing plants were built in Rockland and Boothbay Harbor, in Maine, as well as in Gloucester, Provincetown, and North Truro in Massachusetts. Gloucester schooners were equipped with ammonia freezers to freeze bait caught in Newfoundland[41]—a process that anticipated the factory freezers later developed in Germany and England.

Cold storage, Boston Fish Pier, ca. 1915. Photo: Nathaniel L. Stebbins. Courtesy Historic New England, 23677.

The freezing plant in the Cape Cod dunes of North Truro was part of a small community near an existing icehouse in Pond Village. An intriguing image from 1905 survives, in which the construction built to house the refrigeration machinery has a concrete rational grid and a chimney that dialogues with the pitched roofs of the neighboring wooden boathouses; the ensemble links the beach landscape with the elevated railway that connected Provincetown with the continental network that extended throughout the peninsula dunes. The modest group of buildings supported the fixed weirs that populated Cape Cod's interior coastline.[42] The engineer responsible for the facility operation describes a process targeting nearby fisheries: "The fish are all caught in weirs about one mile from the storage. They are brought in boats to the shore, where they are dressed and washed clean; then they are hoisted to the top of the third story, whence they go down through scuttles into the freezing room, where they are frozen solid."[43] The original construction burned down in 1914 and was rebuilt and expanded with twice its capacity the following year, a massive flat-roofed concrete block that featured in plenty of old North Truro postcards before it was demolished in the 1970s.[44] Another important early fish freezer in Massachusetts was built in Yarmouth Port, at the end of Wharf Lane.[45]

Freezing bait was not freezing food, and there was a general perception that the quality of frozen fish deteriorated when it thawed, "depreciating both in flavor and firmness" and becoming mushy and unappealing.[46] The reasons for this were the long periods of storage—often "two and even three years," after which the fish were "scarcely suitable for the fresh-fish trade"—and the different freezing methods that needed to be adapted according to the physiology of each fish species.[47] There were many techniques for freezing fish, from rudimentary procedures mixing cold water and salt to more complex operations in pans and freezing bins, resulting in either compact boxes of various fish or individual fish frozen by unit. Often an external coating of ice over the fish skin was quickly obtained, supposedly to preserve the moisture of its inner flesh. During the longer freezing process, a breakthrough in freezing technology came with the techniques patented by Clarence Birdseye (1886–1956) and exploited by General Foods in the late 1920s. Birdseye is behind the dissemination of the quick-frozen fillet and the machine to produce them; although he is often portrayed as the genius inventor, his contribution was to systematize a

Cold storage, North
Truro, Cape Cod, ca. 1905.
Courtesy Truro Historical
Society, Cobb Archive.

series of known procedures into an effective production and commercial strategy.[48] Birdseye biographer Mark Kurlansky points out that he was primarily interested in "turning a profit from wildlife"[49]—a focus that was clear in his 1924 patent, the first of many devoted to the "Method of Preserving Piscatorial Products."[50] He designed a procedure to prepare and "ship particular varieties of seafood from a relatively restricted locality that is its natural habitat throughout a radius of land and sea many thousands of miles in extent, and furthermore to hold such foods from periods of plentiful production to times of relative scarcity."[51] The process consisted "in cleaning and dressing the fish ready to cook, packing it in forms to provide independent units, immersing said forms in a refrigerating medium to solidify said units, withdrawing said forms from said medium after their contents have become frozen, removing said frozen units as individual blocks, wrapping said blocks, and packing said blocks in heat-insulated containers for shipping."[52] In 1925 Birdseye moved to Gloucester, where he founded a company, General Seafoods, and bought premises to develop a profitable business from his patent.[53]

Filed in 1927, the patent that would become his fortune was a continuation and perfecting of previous patents, "not limited in its application to any particular class of comestibles or food products" and aiming to "'quick' freez[e] the product into a frozen block, in which the pristine qualities and flavors of the product are retained for a substantial period after the block has been thawed."[54] It was an effective machine that combined the various steps from filleting to packing the product, as well as the specially waxed wrapping paper that acted as an effective protection for the fish during the freezing process, which involved contact with metal plates. Birdseye's machine did not revolutionize frozen fish culture overnight, but it did catch the attention of the Postum Cereal Company, a major food corporation run by Marjorie Merriweather Post (1887–1973), who was in the process of expanding it by acquiring companies specializing in a variety of industrial processes, such as gelatin, chocolate, and coffee. Postum became General Foods, a conglomerate that had the capital, the marketing

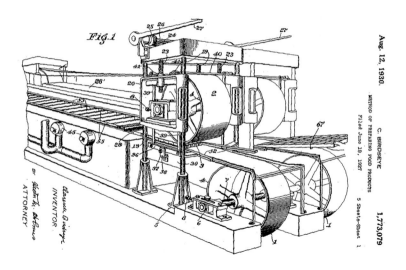

Clarence Birdseye, "Method of Preparing Food Products," patent filed 1927, published August 12, 1930, in Letters Patent, United States Patent Office.

knowledge, and the distribution networks to transform Birdseye's frozen-foods idea into a profitable business.[55] In May 1929, a few months before the New York stock market crash, General Foods acquired Birdseye's "names, patents, patent applications and all assets," transforming the man into a millionaire and a myth.[56] Then, benefiting from the growing cold storage infrastructure that had been developed during the past decades, General Foods expanded frozen fish fillets into a large-scale business.

Birdseye created a machine, not a building. His premises in downtown Gloucester were located by the harbor at the Fort peninsula, occupying a former processing plant for salted and dried cod.[57] He adapted the existing facilities, expanded the constructions, and readjusted their functioning. The technological breakthrough came from the machinery installed inside the building and was not contingent upon the building design itself. However, buildings did change, incorporating refrigeration capacity to obtain better conditions for the machinery to operate in and increasing indoor areas at the expense of large extensions of outdoor drying racks. In Gloucester, the salt fish trade had grown during the nineteenth century on the back of a number of companies, most notably John Pew & Sons, founded in 1849, and Slade Gorton & Company, founded in 1874, the two merging in 1906 as Gorton-Pew Fisheries. The salted and dried cod fish flakes were for a long time part of the Gloucester landscape, registered by the painterly eye of Fitz Henry Lane (1804–1865) and present in many photographs and memorabilia celebrating the town's history.[58]

In addition, Gloucester prides itself on being the birthplace of efficient sailing schooners, its fishing traditions stretching back into the early colonial days.[59] The schooner was a fast boat that achieved a convenient compromise between deck area, hull capacity, and sailing speed, being able to navigate the productive fishing areas, either in the neighboring Georges Bank or farther away to Nova Scotia and the Grand Banks of Newfoundland. The historian Jeffrey Bolster has pointed out how much the ecosystem transformation of the Gulf of Maine resulted from the fishing pressure of coastal communities, the gulf's less saline water benefiting from nutrient-rich flows of oxygenated river waters and its currents benefiting anadromous fish populations, such as alewives, shad, and smelt, to the detriment of pelagic species such as herring and menhaden, the fodder of predators such as cod and haddock. Bolster points out that "cod stocks in coastal waters from Cape Ann to Cape Cod were insufficiently robust to support intensive shore-based fisheries, especially as the process of town-building and farm creation disrupted the habitats necessary to support anadromous fish," and fisherfolk oriented their operations to distant offshore banks.[60] The schooner is an outcome of this offshore occupation, itself a consequence of the ecological impact of coastal fisheries, and a vehicle to support the long-distance fisheries. The focus on dried and salted fish processing, typical of the late-nineteenth-century Gloucester landscape, was the outcome of this complex entanglement between the ecological transformation

153

of local water habitats and the physiological qualities of the main targeted species in distant waters. This landscape of drying fish racks was transformed when steam trawlers, supported by the ice industry and railway systems connecting harbors to cities, started to bring ashore novel species oriented to the then more lucrative fresh-fish market. Birdseye's occupation of Gloucester's Fort benefited from the dwindling of the salted and dried fish businesses (which were contracting after previous expansions, their operations likewise impacted by the relative exhaustion of Nova Scotia and Newfoundland cod populations). Although his target was the burgeoning market of the fresh fish trade, he anchored his business in the disassembling of the salt landscape.

Transforming the fish trade by means of freezing, as would happen in the following decades, meant connecting processing with consumer demand through a cold chain. The cold storage complex and the railway infrastructure provided the large-scale connections, but commercial trade required improved refrigeration power in smaller-scale selling points. The transformation brought about by frozen-fish filleting factories happened in parallel to changes within the domestic space and culture. A major player in the process was the icebox, a device that accompanied the growth of the nineteenth-century ice trade and became pervasive in American households.[61] Iceboxes started being domestic appliances, catered to all budgets, and ranged from modest wooden boxes to sophisticated enameled chests. Far from efficient, their operation relied on the constant supply of ice, which cooled the enclosed space. Temperatures were not stable, and both the drainage of the melted water and the deterioration of its stored goods required continuous maintenance. While the gigantic refrigeration machines that supplied cold houses were being perfected, many inventors and producers thought of reducing their scale and gaining access to the large market of domestic refrigeration. By the early twentieth century, production had started on smaller and more reliable machinery to assist iceboxes and be installed at home, often in basements and storage spaces with pipes connecting to iceboxes. The operation was far from easy: the machinery required frequent maintenance, and electric and gas systems were not as effective as their complicated operation might suggest. The breakthrough came in the late 1920s, when electricity suppliers realized that refrigerators could boost electric consumption and satisfy regular demand from every household. The combined

Drying racks, Gloucester, 1906. Courtesy Library of Congress, Detroit Publishing Company photograph collection.

effort of power companies and appliance producers led to a great improvement in refrigeration systems, incorporating the refrigeration cycle within iceboxes—which have become the refrigerators we are all familiar with today—and within a short period of time, between 1925 and 1938, 50 percent of American households had at least one electric refrigerator in their kitchen.[62]

The ecological impact of this combination of technologies, from the domestic refrigerator to the quick-freezing patents, and all part of a constellation of systems including railway networks and cold storage houses, cannot be overestimated. Fish consumption increased exponentially. In 1932, American frozen fish products weighed approximately 42,000 tons, tripling to 133,000 tons in 1948, while the value of the products kept increasing.[63] As the historian Oscar Anderson set out back in 1953, frozen fillets "permitted the utilization of by-products, the elimination of freight payments on inedible parts, and the marketing of whiting or silver hake, which formerly were used for fertilizer."[64] While expanding its markets to areas where fish was not consumed before and offering food staples that were easy to prepare, the largest achievement of freezing technology was the regulation of natural cycles. Before the establishment of freezing factories, fish prices were constantly fluctuating both in landing operations and consumer trade. Processing fish as frozen products made it possible for stocks to be accumulated: this allowed output to the markets to be regulated to take advantage of low landing costs in productive seasons, while keeping the prices moderate in less productive periods. As happened to the canning industry in the nineteenth century, this capacity to economically tame natural cycles was of great benefit to industrial expansion but at the same time challenging to fish populations.

Scorched earth. When 250,000 Nazi troops retreated from Finnmark, the northernmost region of Norway, in October 1944, they razed everything to ground. On their way, they forced the evacuation of 75,000 inhabitants, setting fire to homes, schools, hospitals, workshops, factories, harbors, and fishing plants.[65] This apocalyptic scenario was the outcome of an occupation that had begun in 1940. In April that year, the German army invaded the country; by June, it had established a Norwegian version of a Nazi government. The occupation, run by a German military directorate with the support of a right-wing puppet government, ruled until May 1945, when the country was liberated, and the Norwegian government returned from its London exile. In the meantime, Germans planned and constructed a variety of infrastructures and facilities, ranging from highways to concentration camps, from nurseries to fish-freezing factories.[66] The incorporation of Norway into the Third Reich had a double goal: it would supply a genetic pool to boost the growth of the Aryan race while at the same time providing access to important natural resources, from precious metals to food supplies. Granting land passage to Murmansk, Norway's Arctic region constituted a strategic military position that could be used to thwart Soviet counteroffensives.[67] But for an important coastal fleet largely consisting of small vessels, the proximity of abundant fishing grounds proved to be of strategic benefit when war prevented both Allied and German fishing fleets from accessing the Atlantic and North Sea fishing grounds. During the war, the Norwegian coast become a privileged source of proteins that could be used to nourish both troops and civilians.

Germany did not have a prewar plan to resource Norwegian fish to feed its population, but the development of fisheries and the establishment of an autarkic policy aimed at national self-sufficiency created the ideal context in which to experiment with such a policy.[68] Although the German fishing industry faced a lack of demand and overproduction, between 1935 and 1938 its trawling catches increased dramatically, from 499,000 tons to 730,000 tons, thus requiring fish to be marketed not only to balance people's diets but also to promote a national product.[69] As historian Ole Sparenberg points out, "Fish was not only promoted as tasty or nutritious food, but Germans were explicitly urged to support their country's economy by eating more fish," with annual average per capita consumption increasing from 11.28 kilograms in 1935 to 13.55 kilograms in 1938.[70] This growth was uneven throughout the country and especially low in inland regions where there was no adequate distribution infrastructure and the local people were not in the habit of eating fish. It might also be the case that the increase in average consumption was linked to army supplies and public canteens. Fish was being used in ersatz products like *Fischwurst* (fish sausage) and *Wiking-Eiweiß* (Viking protein), the latter a powder produced from 1936 onward by a Bremerhaven company to act as a food supplement.[71] In 1939, George Bowling, the anti-hero created by George Orwell in the novel *Coming Up for Air*, described the *Fischwurst* in eloquent terms: "The thing burst in my mouth like a rotten pear. A sort of horrible soft stuff was oozing all over my tongue." Orwell was referring to "these food-factories in Germany where everything's made out of something else ... sausages out of fish, and fish, no doubt, out of something different." Conveying the popular suspicion toward products that disconnected consumers from their food source, Orwell's character commented on the novel fish products: "It gave me the feeling that I'd bitten into the modern world and discovered what it was really made of."[72]

Modern food for a modern world. War was a crucial threshold that marked the entrance of fish fillets into people's consumption habits. Unlike *Fischwurst* and *Wiking-Eiweiß*, fish fillets were still recognizable in relation to their origin: not only did they save on cooking preparation time (one of the downsides that discouraged

overleaf Scorched earth, Berlevåg, Finnmark, December 20, 1944. Courtesy The National Archives of Norway, Riksarkivet, RA/PA-1209/U/Uk/L0224.

157

consumers from eating fish on a daily basis), they also took up less space and were lighter to handle in the distribution process (being traded without bones and inedible parts). British fish and chips prepared the ground for a trade that would separate the edible flesh from the original animal and led to the establishment of a cold-chain network where sub-zero temperatures meant that greater distances were possible between the fishing grounds and the place of consumption. In tandem with ice, trawling—as a form of fishing technology—allowed larger catches to be secured and production scaled up, thus decreasing the price of the final product and expanding popular demand for fish. Hence, in the mid-1930s, deep-freezing fish became a priority for German nutritionists, as the new technology would make it possible to cater to inland populations, balance demand and supply, and source food in distant fishing grounds.[73]

At the time, the reference for frozen fillets were Clarence Birdseye's experiments in Gloucester, over in the United States.[74] His first experiments in 1924 proved successful, and by the early 1930s General Foods had opened up the market to multiple frozen products. In France, too, progress was being made in establishing new patterns of fishing production and consumption—witness the case of the Lorient fishing port and its freezer facility, coupled with freezer depots throughout the railway network.[75] German companies were not far behind. Prior to the war, the former cargo carrier *Hamburg* was converted into a floating processing plant, capable of joining fisheries in distant waters and equipped with mechanical filleting and deep-freezing equipment, allowing the catches of smaller vessels to be processed in situ.[76] Still, there were capital constraints that prevented the small-scale German fishing companies from investing in expensive and risky technology in a market that was already saturated with fish products.

While these modern technology experiments were taking place, cod fisheries in Northern Norway were bound to the long tradition of salted and dried fish. Norway's coastline is defined by the Egga ridge, a 1,000-kilometer underwater mountain chain running from the Spitsbergen islands to the archipelagos of Vesterålen and Lofoten.[77] The Egga is the edge of the Barents Sea, a large basin of relatively shallow shelf and cold polar waters rich in phytoplankton and zooplankton, where capelin, herring, and cod find optimal conditions to thrive.[78] Its cod population benefits from an abundance of nutrients, and young cod, feeding on

160

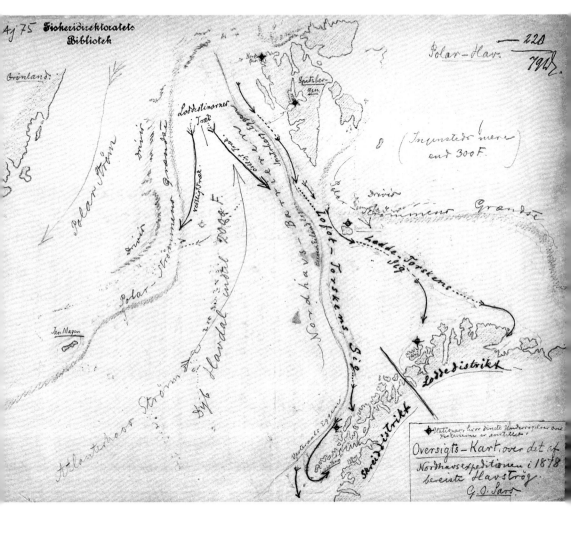

Migration of cod, Barents Sea and Egga Ridge, expedition map by Georg Ossen Sars, 1878.
Courtesy Havforskningsinstituttet, Bergen, IMR 281.

Tørrfisk dryers, Honningsvåg, 1935. Photo: Borgvald Ulvatne. Courtesy Museene for kystkultur og gjenreisning i Finnmark IKS, NO.F001364.

capelin and herring, cruise the basin in patterns defined by the distribution of prey and the water mass temperature. In scientific expeditions between 1876 and 1878, the pioneer oceanographer Georg Ossen Sars (1837–1927) mapped the movements of these fish populations, movements that not only accompany the yearly seasons but also vary according to the fish's age.[79] Sars's observations were later developed and confirmed by further generations of scientists, whose in-depth knowledge supported the consolidation of cod fisheries all along the Norwegian coast. His map explained the reasons for the seasonal fishing habits that were already established and continue to this day. Fisheries in Finnmark start in January, when the mature cod begin to descend along the Egga toward the Lofoten Islands, where they spawn between February and April: this is when the main and most productive fisheries take place. In April/May, younger cod, some three to five years old and reaching 40 to 50 centimeters in length, start following small capelin to the coasts of Finnmark and Murmansk, occasioning another seasonal fishery. These movements correspond to those of the large ocean cod, the *skrei*, while a different population of coastal cod, composed of smaller fish, inhabits the coastal regions, exhibiting a sedentary pattern.

162

There is a perfect match between these ideal seasonal conditions for cod fisheries and the Arctic spring. Strong winds and a
dry environment with temperatures above zero prevent the cod
hung out in the open air from freezing and stop blowflies from laying eggs on them. Norwegian cod production, with its various ups
and downs, can be traced back to the twelfth century: it is a trade
that was initially based on dried fish, called *støckfisk* or *tørrfisk*, and
later on salted and dried fish, such as *klippfisk*.[80] The fine-grained
settlement system of fishing villages was established early during
this process, yet it fell apart and many locations were abandoned
as a result of the European wars of the seventeenth century, with
the fisheries and settlement structure only reestablished in the
nineteenth century. From the 1820s onward, when an important
political shift allowed the privatization of land, cod processing
increased exponentially in parallel with the growth of international export markets.[81] The end of the Napoleonic Wars created
conditions for an important export trade to be resumed, building on a long history of coastal communities combining fisheries
with farming. But unlike their Newfoundland counterparts, which
operated with large vessels and capitalist enterprises, the Norwegian fisheries were coastal ventures supported by small-scale

Tørrfisk dryers,
Honningsvåg, 1935. Photo:
Borgvald Ulvatne. Courtesy
Museene for kystkultur og
gjenreisning i Finnmark
IKS, NO.F001363.

operations and based on a household economy. A large number of fishing communities spread along the coast welcomed irregular influxes of seasonal labor. In the 1950s there were sixty-six of these small communities, a majority of them comprising 200 to 400 inhabitants, and their architecture was characterized by timber constructions settled on natural bays, providing quick access to fishing grounds.[82] Their most distinctive feature were extensive areas of drying racks, where cod were hung by the tail and the catches processed into a lucrative commodity.

Sometimes reaching 9 meters in height, the drying racks of Northern Norway were nothing short of spectacular. These pragmatic wooden structures could be flat (*flathjelle*, which were the convention in the nineteenth century) or gabled with a triangular cross section (*fiskehesje*, first introduced in Finnmark at the beginning of the twentieth century) and covered jagged rocky areas exposed to the winds.[83] The dryers were often hundreds of meters long with high rafters and bracing joists supporting the purlins, from which the fish were hung, one by one, in quantities that defy logic. In spring, each of these monumental constructions stood testament to a massacre of the fish population, and yet their ecological impact seems to have been relatively limited. The fish would remain there drying for weeks at a time, sometimes for months, giving off an unforgettable odor. The smell of the fish was commensurate with the scent of money; geographic and climatic features worked in tandem with the seasonal migration of cod, and the extraction and processing of the natural resource generated considerable income on the international market. This resulted in the continuous expansion of fishing villages up until the 1930s, when the collapse of the international stock markets and the volatility of trading currencies dragged Northern Norwegian fisheries into disarray.

The architects Karl Otto Ellefsen and Tarald Lundevall have researched the urban patterns of these fishing villages in depth, characterizing their architecture and how changes in the sector affected them, from their substantial growth in the early nineteenth century to the present state of global fisheries and their tangled association with tourism.[84] They distinguish four types of traditional fishing villages: (1) organized around a local landowner, the "local king"; (2) seasonal villages, on public land or belonging to various owners and used only during the main fishing season; (3) trading posts, supporting local operations; and

(4) fishing hamlets, with a land structure connected to agricultural production.[85] Considering their fragility and the region's harsh climate, it was not uncommon for these settlements to be abandoned and eventually disappear. As Ellefsen and Lundevall point out, their architecture was "structured for a brief, hectic season, where the objective was to provide mooring space to as many boats as possible, lodgings to a great many fishermen, and enough rack space to dry fish in peak years."[86] Although regional variations existed, there was a striking homogeneity in the architecture of these settlements all along the thousand kilometers of the Northern Norwegian coast. The land constructions, grounded on the uneven rocky coastal landscape, stood over poles similar to the ones used in the wharves. The continuity between land and water, as well as the sinuous adaptation of the endless drying racks to the rocky curves and wind-exposed areas, produced a variety of lively patterns from the same base elements: fish-drying racks; facilities for drying and storing fishing nets and lines; boathouses and warehouses. Living quarters were typically independent architectural forms, often acquired as prefabricated structures, painted white, and strategically located apart from the working areas, where falu red prevailed. A feature of these villages was that both their social structure and built shape differed from agricultural urbanity: there were no distinctive areas for social gathering, no plaza or churchyard.

Berlevåg, Finnmark, ca. 1910. Courtesy The National Archives of Norway, Havnedirektoratet, RA/S-1604/2/U/Ua.

165

Sars's nineteenth-century oceanographic survey was followed, prior to the Great War, by important research conducted by Johan Hjort (1886–1948), at a time when several societies for the promotion of fisheries were being organized in Bergen and the Directorate of Fisheries gained political relevance.[87] While marine ecologists understood the environment and life of fishes, architects and engineers were at work providing infrastructure for their exploitation. The piers, moles, wharves, and other sporadic constructions built in the early twentieth century provide architectural evidence of the centralized organization of northern fisheries. The Statens Havnevesen, or port authority, conducted a large program for small infrastructures that consolidated the fragile settlements scattered in islands and remote places. These works were documented in a comprehensive photographic survey that exposes the artificialization of the coast, creating harbors where larger boats could be safely moored. Looking back at these freshly built moles, we can see an intuitive sense of form: snaking figures no more than two or three meters in height dialogue with the shore's irregular rocky patterns, appearing and disappearing between the mirror-like surfaces on the harbor side and the rippling water crashing on the boulders that protected it.

Nyksund stands out among these operations. Located in a western tip of the Vesterålen archipelago, two tiny islands were bridged in 1877 to create a protected harbor, 45 by 220 meters in size. The singular characteristic of this basin is that it is ringed by a wooden two-story deck, which provides easy access to the water below and level entry to the ground floor of the surrounding buildings. Most striking are the four large multipurpose warehouses, all similar in construction, erected in concrete on the north side of the harbor. Their wooden façades with overhanging cranes detached from the gabled roof, mimicking traditional buildings, belie the technology deployed in their construction. The rusticated cement decoration on their east-west façades reveals their true nature: they are not part of an adaptive settlement located around a natural harbor but the product of a significant capital investment, remotely designed and geared to exogenous building technologies. This becomes even more evident in the harbor's southern wharf, where a wooden warehouse built in 1871 was later expanded with a freezer that stands out as a geometric abstract block of concrete. Although Nyksund's

166

P.52-1.

Breakwater under construction, Valvær, ca. 1900–1910. Courtesy The National Archives of Norway, Havnedirektoratet, RA/S-1604/2/U/Ua.

Nyksund, ca. 1900. Courtesy The National Archives of Norway,
Vesteraalens Dampskibsselskab, RA/PA-1189/U/Ue/L0002.

location is extreme, formerly accessible only by boat on a tip of the Vesterålen island, its forms were conceived abstractly by technicians at drafting tables. Their aim was to settle fishers as close as possible to the spring cod runs.

Cod prefer eating capelin to herring.[88] Whereas capelin is of little commercial value, herring is one of the most sought-after species for human consumption. Although the ecosystems of these species are intertwined, herring has a higher critical temperature threshold than cod, meaning its most important spawning grounds can be found to the south, in western Norway, a region rarely reached by cod.[89] Nonetheless, when it comes to understanding the development of freezing in Norway, the two are intertwined, reflecting the centralized logic of political decisions. For instance, it was in Melbu, in the south of the Vesterålen at the heart of the spring-spawning cod region, that in 1887 the first company used steamers for fishing, targeting herring, not cod.[90] Meanwhile, the management of the herring trade during the Great War prompted the construction of a large freezer in Ålesund in 1918.[91]

The war disrupted previous trade patterns, and currency fluctuations and protectionist measures combined to bring further difficulties to the Norwegian fish exports, a problem exacerbated by the collapse of world commercial networks after the 1929 stock market crash.[92] In 1931, Norway had twenty-seven refrigeration or freezing facilities,[93] most of them in the south, and the technological advantages of cold were at the forefront of the interwar debate on how to modernize the national industry. In the northern region of Finnmark, several producers engaged in the export of iced boxes to England and Germany. Natural ice was ground and compacted in wooden boxes containing fresh fish, then shipped to its destination markets with a view to racking up large profits. Up to 70,000 crates were exported in 1936, a business that many saw as "pure gambling," since the product often did not reach its destination in a suitable condition.[94] The fact that in 1931 a Melbu producer advertised its frozen products in Chicago exemplifies the ambitions of the trade and demonstrates a commercial awareness that there was a need to compete in new markets. However, 90 percent of cod landings were still being salted and dried. It was in this context that a commission was organized, leading to parliamentary approval of the 1932 national plan to implement freezing capacity.

The plan, and the political debate surrounding it, played a fundamental role in conveying a sense of the economic advantages of transforming the culture of cod processing. In its preliminary report, the fish commission presented a table comparing the yields and profits of several countries between 1926 and 1928.[95] The table told a compelling story: Norway lagged behind all its competitors in the price per kilo of the fish it sold, which averaged 0.09 kroner, half the German rate of 0.20 kroner and almost a quarter of the 0.35 kroner charged by the British and Dutch. And yet its yields were more than twice those of the British. How could so many fish generate so little profit? There were several reasons for the disparity, but a visible one was the wealth of urban markets for fresh fish. Simply put, transforming cod into *tørrfisk* or *klippfisk* meant decreasing its value. Hence, the 1932 plan devised a strategy to develop a cold chain and modernize its fishing culture. As the committee stated: "It is a matter of the future, perhaps of vital importance for our fisheries, that the catch can be frozen."[96]

To transform the Norwegian fisheries landscape, the administration designed an eloquent map: 212 fishing villages were listed and connected through a logical network of fishing production, including large, medium, and small fisheries.[97] It included a hierarchy whereby red-underlined names were attributed to the main freezing houses and had a key position in the comprehensive plan. Twenty existing cooling facilities were to be integrated in a network of fifty-six new buildings, inscribing the remote northern areas of Finnmark within a national system capable of balancing production through the seasons and regulating prices on the basis of stable storage and trade. Like every modern plan, it required time to be implemented, and most of the buildings got no further than the paper they were drawn on. More important, the shift from traditional to modern fish processing did not happen: in 1939, only 600 tons of cod were being frozen, representing 0.18 percent of the total catch.[98] What rescued Norwegian fisherfolk were not the freezers but the Raw Fish Act. Passed in 1938, it established a legal framework for granting fishers' organizations and cooperatives the right to establish minimum sale prices, thus empowering fisherfolk and securing a balanced relation between the primary workforce and fish buyers.[99]

Frozen fish required a market and production knowledge, and this was brought to Norway by German companies, operating in consortium with local fish-processing plants. The first to arrive was Nordsee AG. Established in 1896, Nordsee was one of the few German companies that by the 1930s possessed a trawling fleet,

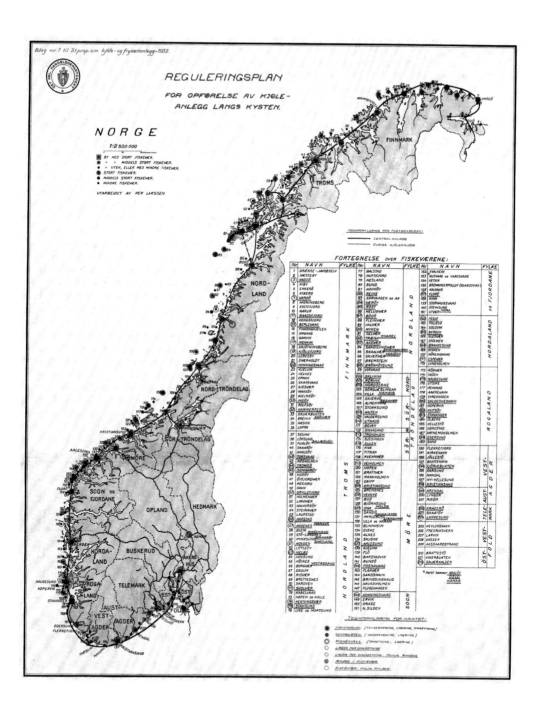

National plan for cooling and freezing facilities, Norway, 1932. Courtesy Stortingarkivet.

fish-processing plants, and a distribution network throughout Germany. In 1937, the Dutch group Unilever became Nordsee's main shareholder, and in 1939 Solo Frost GmbH was created to manage the company's freezing activities, ranging from fish to fruits and vegetables.[100] Just before the German invasion of Norway, Nordsee acquired a license for Birdseye plate-freezing technology and initiated the construction of a freezing plant in Trondheim.[101] It was followed by a second plant in Bodø, while the German consortium Vereinigte Tiefkühlgesellschaften Lohmann started operations in Hammerfest (before moving to Svolvær in 1944).[102] The fourth German plant in Norway was located in Melbu and run by the local entrepreneur Gunnar Frederiksen in collaboration with the Hamburg company Andersen & Co. The Melbu factory became a reference in the history of Norwegian freezing culture, having begun operations after an Allied bombing raid in Svolvær in March 1941 that sank the *Hamburg* factory boat, whose filleting equipment was salvaged and installed in Melbu.

Advertising photograph for domestic refrigerator with freezing compartment, Lykkepotten, 1955. Photo: Kjell Lynau. Courtesy Riksarkivet, Billedbladet, RA/PA-0797/U/Ua/L0042/0652.

The Nazi factories in Norway were a game changer. They increased production exponentially, from an average capacity of 10 tons per day in 1939 to 350 tons per day in 1943. To achieve this output, a large industrial workforce was required. In Bodø, the fish plant employed about 1,100 workers, including 100 Germans, 120 Norwegians, and at least 760 war prisoners.[103] Forced labor was key to transforming the industry, and the increased production guaranteed the flow of catches harvested by fisherfolk who risked their lives working in a war zone and the mined waters of the Atlantic. Fish buyers adapted their activity and, instead of drying fish for a disrupted international market, iced and packed the catches to supply the growing freezing industry. Historians agree that the Nazi factories brought the knowledge and experience necessary to radically transform Norwegian fishing activity in the postwar period.[104] Even while the war was still in progress, both the Nazi administration and the Norwegian government-in-exile started giving thought to postwar Norwegian fisheries based on frozen products instead of the prewar processes of salting and drying.

More important than the factories in precipitating this shift was the scorched earth policy carried out by the Nazis when they retreated from Finnmark. The level of destruction was so overwhelming that everything had to be restarted from scratch, an unexpected tabula rasa upon which to implement the ideas planned back in the early 1930s. The reconstruction process was narrated from an urban perspective by the architect Trond Dancke (1915–2001), who moved to Vardø in 1945 and was a major contributor to the building effort.[105] The goal was clear: "The recovery was not to be the reconstruction of what had been burned. Everywhere there was an element of new construction, both in quality and quantity."[106] The office focused on town planning, housing, and amenities such as schools, hospitals, and churches.[107] It inherited the tasks of the Brente Steders Regulering (BSR), a technical structure for the reconstruction of war-damaged areas. After liberation, its leading architect was Erik Rolfsen (1905–1992), who replaced Sverre Pedersen (1882–1971) after the latter was proscribed for collaboration with the Nazi government.[108] Many architects were involved for short periods in the work of northern reconstruction, and one of the distinctive features of BSR in the north was the role of women architects, who found the emergency situation to be a meaningful locus of professional practice.[109] Although the reconstruction of the fishing industry was conducted by the Fiskeridirektoratet in Bergen, there was a connection between the urban strategies deployed by the BSR and many of the design features, notably the novel freezing facilities developed by the fisheries department. The design of buildings to support fisheries was a constitutive element of the resurgence of the various communities, and standard plans for the layout of wharves, warehouses, freezing facilities, and drying racks were defined according to new regulations and in contrast to traditional local building practices.[110] A 1947 perspectival drawing characterized by the distinctive lines of Bitte (Dina) Bergh Sewell shows Vadsø harbor with a modernized look, the fishing setting accompanied by high-rise dwellings in the background.[111] The vernacular flavor of the communities was replaced by the modern look of a welfare society.

In 1948, architects met in Alta, in Finnmark, to discuss reconstruction.[112] Dancke presented the layout for an "ideal" fishing village, with multiple functions assigned to clearly identified zones: the harbor and the basin; the piers and the fish plants; a service station for storage and cold chambers for bait; drying racks; residential areas; and the town center. Observing the work of the reconstruction architects, Ellefsen and Lundevall emphasize how the establishment of community centers and social areas independent from the harbor derived from modern planning culture. In his book, Dancke included an instructive comparison: designs developed for Finnmark were juxtaposed with works

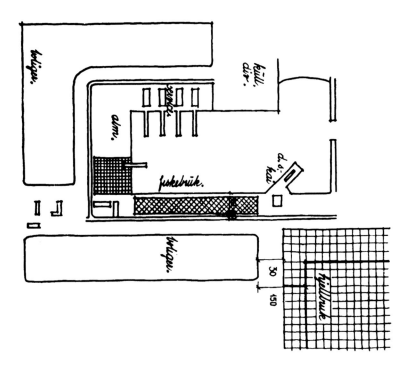

from Pompeii, Priene, Venice, and other urban references from the European canon.[113] Besides the orthogonal ideal fishing village, the reconstruction offices also provided reference drawings for modern facilities to be adopted for the multiple functions associated with fisheries.

Reconstruction of Finnmark was quicker than anyone had expected, both in fish landings and resettlement.[114] Despite the tendency to concentrate population, most settlements were rebuilt in their previous locations, even if they were abandoned soon after. By 1948, the total weight of fish landings was already equivalent to prewar levels. The major difference was the number of fisherfolk, which decreased from 9,800 in 1938 to 6,930 in 1948, and their secondary occupation, which shifted from agriculture to construction.[115] But the major transformation that had an impact on both the ecosystems and the landscape of frozen fillets architecture was the strength of the cooperative movement and the governmental support ensuring that fisherfolk would own their boats. It was a political choice, which separated fisheries, processing, and trade and granted fisherfolk the means to be self-employed. The main fishing methods continued to be tubbing, gillnetting, and jigging, with a significant increase in purse seining

174

ADSØ BYPLAN
RSPEKTIV MOT FISKETORGET
OSØ 26 JUNI 1947

in the 1950s.[116] Instead of the major transformations that could be expected on a tabula rasa, an injection of capital preserved the prewar modus operandi while laying out the basis for the major technical and cultural transformations that frozen fillets were part of. The small investment required for natural drying meant that the region quickly returned to fisheries attuned to prewar culture.

Meanwhile, politicians, economists, and architects were planning a different future for the modernized northern fishing settlements. Following the ideas outlined in the 1932 plan to shift fish exports from natural processed to frozen fish and profiting from the technological advances bequeathed by the German factories, the government conceived of a state-owned company to foster a progressive shift from traditional fish processing to what was expected to become the more profitable business.[117] Finnmark was chosen as the site of the experiment, owing not only to the reconstruction work being carried out there but also to its longer fishing season, which guaranteed a year-round supply of resources for its industrial processing. It was in this context that the state company called Finotro, the "fishing industry of Finnmark and Nord-Troms," was dreamed up and implemented.[118] The idea was to transform preindustrial Northern Norway into a

Bitte (Dina) Bergh Sewell, perspective of planned fish market, Vadsø, 1947. Courtesy Vadsø municipality.

modern society, with year-round secure jobs, a balanced population structure, and education and health services. The abundance of spring-spawning cod runs was supposed to nurture this society in the making. This gave rise to the state-funded scheme that placed Finotro between the fisherfolk's cooperative structures—which would supply the resources—and the national export company, Frionor, which oversaw the branding and commercialization of modern fisheries output.

Between 1951 and 1956, of the twelve originally planned factories, Finotro built seven facilities in Finnmark, including two larger buildings in Båtsfjord (opened in the autumn of 1951, with a production capacity of 20 tons per day) and Honningsvåg (opened in March 1952; 15 tons per day), the latter hosting the company's headquarters. The Vardø factory (10 tons per day) went into operation in July 1952, followed by Berlevåg in 1954, Kjøllefjord in 1955, and Mehamn and Skjervøy in 1956 (5 tons per day each). Unlike the reconstruction projects relating to housing and public amenities, most of them built in wood, these were concrete flat-roofed structures, whitewashed boxes standing several stories high and hovering over the water. The large square windows illuminated the interior spaces, where rows of women worked on sophisticated stainless-steel machines to process the fish, and the direct access to the quay ensured functional and immediate communication between the boat and the factory. Their design was overseen by the directorate of fisheries in Bergen and conducted by the engineer Hans Tveitsme (1916–2002), who was also a member of Finotro's administration board.[119] A series of 1952 aerial pictures of the Honningsvåg factory show the newly built structure served by roads that were freshly carved from the hillside, making it accessible by car. With large square windows to cast light into the interior, the freezing factory contrasts with its neighboring building at the other end of the quay, whose design matches the depictions of fishing facilities in Dancke's account: the gable-roofed warehouse combines two perpendicular overlapping volumes built in wood and covered with corrugated iron sheets. The two buildings highlight the differences in the way the BSR architects and the engineers of the fisheries department viewed the same modernization process.

A freezer is not necessarily a concrete white box, and Norwegian decision makers were aware of it. In an internal document, a senior engineer remarked on the design strategies for Finotro's facilities:

> A concrete block must perhaps be considered necessary and expedient for cold storage plants, but fish processing factories and other production facilities should be placed in cheaper buildings that may be modified and adapted to current developments, at least when there is space. The fish processing industry is continually evolving, and no one can say whether a facility that is functional today will continue to be so in 5 years. Large concrete buildings severely restrict freedom of movement, and it is typical of the American fishing industry—and of American industry in general— that buildings are mostly light and virtually enclose the equipment as a flexible holster that can easily be modified according to how this equipment changes.[120]

Yet concrete prevailed, and it was the contrasting linear volumes of the freezers that defined the modern fillet landscape promoted by the state in the reconstruction of Northern Norway. As for the symbolism of its architecture, the choice to promote the frozen fillet industry was a political choice.

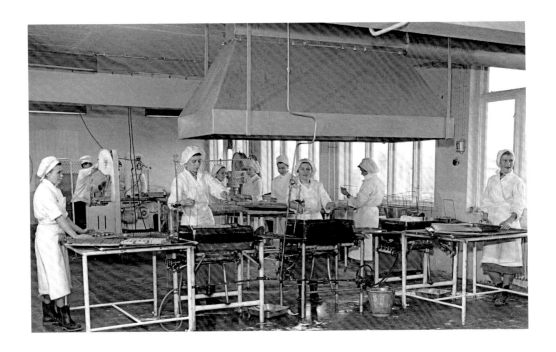

A striking feature of the Finotro business model—and a matter of internal disagreement in its policies—was that it combined the production of frozen fillets with salted and dried cod. This hybrid output can be understood in terms of the company's initial failure to conquer the markets. While its optimistic expectations originally focused on the large, expanding US market, in reality most of its exports in the first decade were destined to go to Italy, Austria, and France.[121] Work soon needed to be done in the domestic market to find a home for the innovative fish products and transform the unpleasant associations of frozen products with the Nazi occupation.[122] Another hindrance was the relative stability in the value of frozen products, set against the added value of salted and dried cod.[123] Between 1951 and 1955, while the value of frozen exports rose from 2.05 to 2.93 kroner per kilogram, traditional exports for Italy rose from 3.87 to 4.80 kroner, meaning the profits of naturally processed fish, with less incorporated value, were faring better than frozen fillets, whose production was more expensive. Hence, it is not surprising that long rows of drying racks continued to proliferate within the newly built Finnmark landscapes.

In fact, beyond its revolutionizing of fish processing, Finotro operated as a buyer for the traditional coastal fisheries. The

1938 law stipulated that only fisherfolk could own fishing boats, preventing trawling from escalating, with the result that fishing remained concentrated in coastal waters. Besides the tricky political and commercial zigzags that secured the functioning of a system that was far from perfect, relying on coastal fisheries meant that the factories ran below their ideal production levels, operating between 100 and 140 days per year, despite having been planned to operate for over 200 days. A major reason was the lack of capacity of coastal fisheries to supply the larger quantities of resource required to reach full production capacity and overcome financial losses. Flying in the face of its own strategy, the government approved a vertically integrated company in the neighboring village of Hammerfest: the company operated a freezing factory while also possessing its own trawling fleet.

Bendix Heide, a salt fish trading company from Kristiansund in the south of Norway, rented a facility built in 1949 in the razed grounds of Hammerfest to initiate the production of frozen fillets. In 1952, the company was joined by Freia, an important chocolate producer from Oslo. Freia had the political leverage to secure the controversial trawling permits that, in many ways, undermined the policy for coastal fisheries that Finotro was promoting.[124] The

Weighing and packaging section, Findus, Hammerfest, 1955. From the photo album "Eventyr i ishavet," courtesy Museene for kystkultur og gjenreisning i Finnmark IKS, GMH.F004365.

vertical integration of fishing and production, and the access to its own trawling fleet, granted the Hammerfest company strategic advantages, both in the cost of raw materials and access to productive offshore fishing grounds. In the meantime, Freia had acquired a Swedish company that owned the label Findus, which became the name of the Hammerfest company.[125] The mergers within the international food trade became even more explicit when, in 1962, Findus was purchased by Nestlé, becoming a global brand. These commercial alliances with chocolate companies brought fish to the heart of the food business and were comparable to the assimilation process that saw Birdseye become part of the Postum Cereal company and General Foods. With Nestlé in the picture, Norwegian frozen fish exports skyrocketed, from 29,000 tons in 1961 to 117,000 tons in 1969.[126]

Unlike the frozen caverns of American cold storage facilities, the atmosphere of Findus was that of a light, bright-white, sanitized space. Contrasting with the vernacular tones and falu red of the wooden buildings of previous fisheries, the rational forms of the concrete structure where whitewashed, clad with white tiles, and populated by women dressed in white smoke frocks and rubber galoshes. Large windows and light, reflected in the glazed surfaces of the wet floors, were a main feature of the new architecture, in which machinery was the main actor. Promotional films from that time show diligent administrators checking the sequencing of the mechanical procedures the fish goes through once it has been offloaded on the quays, as its viscera are transformed into oil and fishmeal and its flesh into shiny white fillets.[127] The factory also became a gigantic kitchen, with fillets transformed into fish sticks, battered, and then frozen ready for consumption. If hygiene was a topic in cold storages, the architecture of frozen food factories elevated cleanliness and efficiency to a new level of obsession, as food safety was the key for its commercial success.

Would such transformation in Norwegian cod landscapes have been possible without the desolation left behind by the Nazis? Hundreds of kilometers south of Hammerfest, the Vesterålen example of Myre examined by Ellefsen and Lundvall demonstrates a parallel process of transformation that was not preceded by destruction. In 1947, Gunnar Klo moved its premises from the coastal harbor of Nyksund to Myre, while the one fisherfolk's cooperative settled on the opposite side of the bay, below the Sommarøy hill. Over the years, the site's logistical advantages resulted in

Nyksund being abandoned, as its harbor was too small for larger vessels, the quays too short for larger buildings, and its position too dependent on sea transit for exports. In Myre, the Øksnes Langenes Fiskeindustri (ØLF) incorporated the fisher's cooperative into one vertically integrated production chain and obtained several trawling permits. The Fryseriet freezer plant opened in 1954, likewise designed by the Bergen department of fisheries, probably with the participation of Tveitsme.[128] As the freezing industry became the dominant force in Norwegian fisheries after the war, Nyksund experienced the same fate as other fishing villages when motorized vessels replaced rowing and sailing: the former settlement was abandoned in favor of a more strategic location, in this case Myre.[129] Both the abandonment of the old site and the mixed landscape combining traditional and modern fish processing suggest that the destruction left behind by the Nazis was not the basis for the growth of the frozen fish industry. Finnmark had been laid waste while the islands of Lofoten and Vesterålen had been relatively spared, yet both were affected by an active state policy that fostered the transformation of the villages' social structure, moving from the prewar network of fisherfolk dependent on powerful

Fryseriet, Øksnes Langenes Fiskeindustri, Myre, ca. 1955. Courtesy Museene for kystkultur og gjenreisning i Finnmark IKS, GMH. F004365.

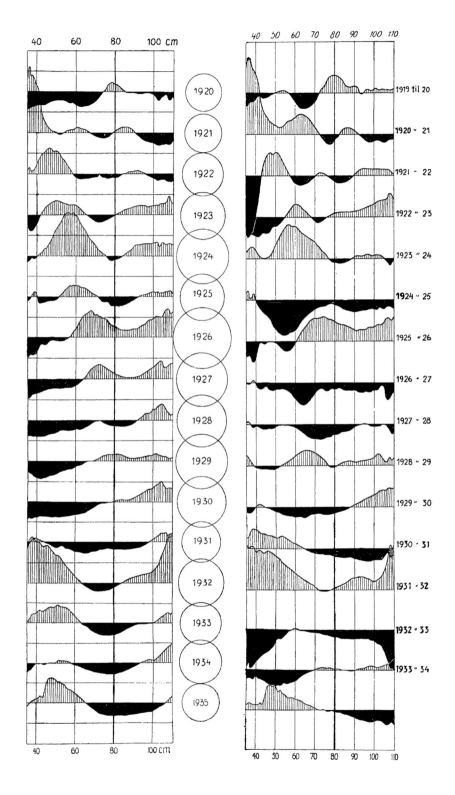

40   60   80   100 cm

1920
1921
1922
1923
1924
1925
1926
1927
1928
1929
1930
1931
1932
1933
1934
1935

40   60   80   100 cm

40 50 60 70 80 90 100 110

1919 til 20
1920 " 21
1921 - 22
1922 " 23
1923 " 24
1924 " 25
1925 " 26
1926 - 27
1927 - 28
1928 " 29
1929 " 30
1930 · 31
1931 · 32
1932 " 33
1933 " 34

40 50 60 70 80 90 100 110

fish buyers to a postwar solidary structure involving fishing, processing, and trade. The three stages of the fishing economy were part of a welfare society in which the state provided education, health, and work for everyone. The policy was underpinned in technical terms by the freezing industry and supported, in turn, by the abundance of fish resources. Trawling allowed the strategy to evolve between the 1950s and the late 1970s, when the fishing grounds started to show signs of exhaustion.

A 1950s drawing published by Dancke representing data obtained from the fisheries department shows Finnmark fishing areas for cod, halibut, haddock, pollock, and herring.[130] The 12-mile line bounding a protected fishing zone—a topic that was becoming a focus of international diplomatic discussion—defined an interior area where coastal fisheries could haul a significant diversity of species.[131] Yet, offshore, Dancke depicts several banks of productive fishing grounds for cod, halibut, and haddock. Before World War II, these fishing grounds were being exploited by British, German, and Soviet fishing fleets, and although Norway saw a large increase in offshore catches after 1953, these fishing banks continued to be explored by growing numbers of British and Soviet vessels. Most of these fish populations were transformed into frozen products.

In 1962, a decade after the simultaneous leap of Finotro, Findus, and ØLF into freezing processing, Norway's production of frozen fish exceeded *støckfisk* for the first time.[132] At that moment, fisheries scientists documented "the story of an old and famous fishery being drastically reduced by the increased competition for the riches of the sea."[133] The modern fisheries, combining larger vessels and diesel power, boosted by the use of synthetic nets whose "fishing power" was "from three to four times greater than the cotton or hemp nets," showed a systematic decrease in mean annual catches per unit of fishing effort.[134] Although catches were increasing, they required more effort and were becoming less profitable. More significantly, the maturity of the spawning cod captured in the Lofoten region was decreasing, from an average age of ten years in the period from 1932 to 1957, and a peak of 11.5 in the years after the war, down to less than nine in 1956.[135] A diminishing mean spawning age was a clear sign of overfishing. While observing a "striking reduction of the mean age" happening "simultaneously with a serious reduction in the abundance" of the spawning cod population ("but not of the young fish"), in

facing Fluctuations of fish size and catch "per man per day at sea" of Norwegian cod fisheries. Graphs by Oscar Sund, in "The Fluctuations in the European Stocks of Cod," *Rapports et procès-verbaux des réunions* 101 (1936): 7.

1964 the scientists Gunnars Sætersdal and Arvid Hylen flagged the possibility "that a physiological change in the maturity age has occurred."[136] Regulations such as determining mesh sizes to save smaller fish and establishing international maritime borders—excluding most international fisheries from Norwegian and Soviet waters—avoided the impending ecological disaster, managing a significant decrease in mortality rates. It was not until the 1990s that the Arctic cod population reached its lowest recorded levels.[137]

By assessing long-term records of fish populations and landings, scientists recognized the entangled connections between environmental and human factors in the variation of Arctic cod maturity and weight.[138] The large scale of its population meant it was only with the postwar dissemination of trawling and fisheries supported by freezing technology that the mortality rate started to affect the Arctic cod population's health and stability. Unlike the infamous case of the Grand Banks, which was exhausted to the point of no return, Norwegian fisheries managed to go through the bottleneck of unregulated predatorial catches and recover a relative strength, with large specimens being regularly caught. Nonetheless, the postwar transformation from dryers to freezers had a significant impact on the abundance and functioning of cod populations. The freezer and the cold chain, from the American cold storage to the Norwegian filleting factory, were the architectural typologies that supported the ecological transformation of marine environments.

805 MILLIONER TORSKEHOVEDER, LOFOTEN.   K. KNUDSEN. BERGEN.

5   *The Lugger and the City*

A naval squadron anchored in the river is a means to invest the city with symbolism. Each year between 1936 and 1973, without exception, the Portuguese dictatorship choreographed an extravaganza in the first days of April, involving thousands of extras, to promote the status of cod as a source of patriotic cohesion.[1] A photo taken in Lisbon one Sunday morning, on April 4, 1965, displays just such a strategy: at the center of the image, the Portuguese Renaissance masterpiece of Belém Tower accompanies the cod-fishing fleet by the river, framing the operation within a monumental triangle created by the tower, the Jerónimos Monastery, and the Chapel of Saint Jerome. The protagonist in the photo is not the fish but an archbishop, carrying his miter and crosier, granting benediction to the fleet in preparation for its departure. As divine protection is summoned for the fisherfolk, their families in the city follow the sacred moment. In the opposite direction, a group of reporters hold their photo and television cameras to provide media coverage for the event, recording the characters and the protagonist. From behind, we can recognize without difficulty the figure of Admiral Henrique Tenreiro, the "boss" of Portugal's fisheries.[2] The lugger, a sturdy fishing boat derived from the

overleaf Cod fishing fleet benediction, Belém, Lisbon, April 1941. Courtesy Biblioteca Central de Marinha/Arquivo Histórico, FG/009/28/1-11.

189

fast American schooner, was the hinge between the cod and the city, echoing the relations between the marine ecosystem being exploited and the politics supporting this exploitation, a dynamic that propelled the transformation of building types such as drying racks and entire landscapes. The scene of anchored luggers and mobile politicians is a powerful image showing one fish's impact on architecture and the city.

That Sunday, a morning newspaper cover recalled the historical dimension of that "lovely location": "Many of our sailors were there, and there they prayed before setting sail on their seafaring adventures, in conquest of other lands."[3] Today, we can still see in Lisbon the built remains of these outdoor masses, which mobilized fisherfolk from all over the country and enabled heads of state to inscribe the activity of cod fisheries in Portugal's collective imaginary. The benediction ceremony happened for the first time in 1936, as part of the Campanha do Bacalhau: this cod campaign paralleled the country's wheat campaign, an emulation of the Italian model, the Battaglia del Grano.[4] In 1937, while the Spanish Civil War was in full swing, there was intense political instability, and Portuguese fisherfolk threatened to refuse to board ship for the cod-fishing campaign. They demanded safer working conditions and were suspicious of the state-driven corporative reorganization of fisheries that was taking place.[5] Tenreiro opposed the celebration of national identity to the fisherfolk's struggle and, after suppressing the uprising, transformed the benediction into a magnanimous media parade dramatizing the strategic importance of fisheries: whoever challenged the cod campaign would be challenging the very foundations of the nation, the homeland of brave sailors. The admiral, who was experienced in organizing fascist parades, used the same scenography techniques with all the pomp in the blessings of 1937 and 1938, under the patronage of Lisbon's cardinal-patriarch.[6] In the following years, almost every celebration was conducted by the archbishop, who, as the Sunday newspaper reminds us, was "also connected to the life of the sea, his father being a fisher."[7] The masses culminated in a reception by the head of state, the dictator António de Oliveira Salazar (1889–1970), who, in 1965, "hosted the fisherfolk with the utmost cordiality, exchanging ideas with them while offering them a glass of port" and some almonds.[8]

In this chapter, we will follow the Portuguese luggers to the Grand Banks and the landscape of cured cod on the coast of

Europe, moving from the changes in fishing vessels and gear to onshore processing facilities. Although their viewpoint was rooted in economics, early twentieth-century political strategists understood the ecological differences between cod and sardine and designed their policies accordingly. As a result, the characteristics of the fishing fleet ended up determining the landscapes in which cod were cured, including the strategic position of cold storages, which were built on a monumental scale to both affirm the policies and exert control over commercial networks. Eventually, cured cod harvested in distant waters was preferred to refrigerated or frozen species caught closer to home. Prior to the twentieth century, cured cod was a staple food, not one that people enjoyed eating. It nevertheless became a vehicle for forging a sense of national identity despite its reputation as a low-quality product, its varied provenance, and the economics of harvesting it. Priority was ultimately given to replacing imported fish with Portuguese catches so as to remedy trade imbalances, even as changes gradually took place in marine ecosystems. These ups and downs ended up establishing a rhetoric that had little to do with the cultural and ecological transformations that were then

Cod fishing fleet benediction, Lisbon, April 4, 1965. Courtesy Biblioteca Central de Marinha/Arquivo Histórico, FG/9-28-2-4.

193

happening, an outlandish history that emphasizes the environmental subtleties between land and sea.

Until the late 1950s, most of the Portuguese cod-fishing fleet consisted of luggers, sailing vessels with three or four masts derived from the schooner, capable of combining relatively quick navigation with a large cargo hold. Schooners and luggers are emblematic of North Atlantic sailing fisheries in the nineteenth century, and their success was directly related to dwindling fish populations. The history of fisheries is one of continuous technological progress designed to overcome an ongoing decrease in catches. When a fishing ground starts to be trawled, catches tend to decrease, and in order to balance profits, fisherfolk increase the pressure on the animal by adopting different techniques. Two compensation principles are classical signs of overfishing: technological innovation, coupled with the ability to extract more fish; and territorial dispersion, as new grounds are sought for extraction. In both cases, the immediate success of the solution found to circumvent the decreasing catch masks the long-term failure of the strategies adopted.[9]

Lugger *Hortense*, Grand Banks, 1953. Photo: Eduardo Lopes. Courtesy Centro Português de Fotografia, EL/306.

In the cod fisheries above the Grand Banks off Newfoundland, one such major innovation was introduced by French fisherfolk between 1787 and 1789: the *ligne dormante*.[10] Literally translated as the "sleeping line," or setline, it consisted of a long line with secondary lines holding one hook each, and it quickly replaced the hand lines used individually from the main boat. Since cod is a demersal fish, feeding from smaller fish, crustaceans, and mollusks from the seabed, the setline was highly effective as it distributed bait across a broad area and increased the number of catches made by individual fisherfolk. The technique was widely adopted between 1810 and 1820 when, after the Napoleonic Wars, French fisherfolk set them from shallops launched from a main ship.[11] In parallel, small, flat-bottomed boats with low gunwales started to be used off the East Coast of the United States, notably in Gloucester, where cod was still being fished off Cape Ann. The wherry, or dory, started to be used from the shore and, from the 1830s on, to be launched from schooners.[12] Setlines were adapted by the Massachusetts halibut fisherfolk, who used a tub to roll them, thus the name tub-trawling.[13] The setline, or trawl, was launched by hand from the dories that sailed off the main schooner to enlarge the fishing area covered by each boat. As opposed to the French shallops, which were hard to maneuver and prone to capsize with several fisherfolk on board, dories were tiny, lightweight boats with only two fisherfolk (the Portuguese used to sail alone), their flat bottoms allowing them to be stacked on the schooner's decks. The cod populations off the United States had dwindled by the 1830s, whereupon this technique, as well as the schooners and luggers, started to spread beyond the Grand Banks, reaching the shores of Labrador in the mid-nineteenth century.

In Portugal, far away from the natural habitat of cod, the first attempts to develop these fishing techniques occurred in 1836.[14] The investment in long-distance cod fisheries, a few years after the end of the Napoleonic Wars, is associated with an 1830 legislative decree abolishing "all rights, contributions, tithes, duties, or taxes ... that have hitherto been levied or demanded for fish caught by Portuguese boats ... , including all the fish caught ... in distant fisheries."[15] In an 1835 booklet, which was intended to facilitate the establishment of a company in Lisbon, a Portuguese ambassador in Washington published a description of how cod was being fished and processed in the United States. He carefully describes the international treaties defining the territorial ownerships of fisheries, underlining that "in any case, cod is a fish that is so abundant in the locations mentioned that our fisherfolk have a large area in which to search for it, wherever is most convenient."[16] The description is technical, designed to elucidate how the fishing gear and boats should be operated in accordance with the international politics of fisheries and sea borders. A century later, the lugger fleet being blessed in Lisbon was not much different from the ambassador's description.[17] Cod fisheries were by no means an "immemorial practice" and certainly not part of the Renaissance sailing culture, as proclaimed by the nationalist propaganda. They were a system developed on the American coast, incorporating French techniques, which spread offshore to compensate the effects of inshore overfishing.

Long-distance cod fisheries were brutal. In spring, a ship would sail from Europe laden with salt and a crew of thirty to fifty fisherfolk, equipped with an equivalent number of dories. Once in the Grand Banks, fisherfolk would set out from the lugger in their dories each day and launch a setline carrying hundreds of baited hooks, heading back to the main boat to unload the catch when the dories were full. On deck, fisherfolk resumed work in a production chain where the fish would be cleaned, cut, and split to be salted and packed in the hold. The season would last about

Dories arriving to unload daily catch at lugger *Gazela*, 1953. Photo: Eduardo Lopes. Courtesy Centro Português de Fotografia, EL/234.

five to six months, with rare visits to the Newfoundland ports. Once holds were filled, the boat would sail back to Europe. Trawling was also rough, yet voyages would last half of the time before returning to Europe with approximately the same cargo. Crews were smaller, the ships larger, steel hulled, and motorized. And the main difference was that fisherfolk would not leave the deck, from where the side trawl was hauled, thus avoiding the plethora of incidents involving tiny dories sailing alone in the often foggy and unpredictable open ocean.

In the 1930s, steam- and diesel-powered trawling vessels were not a novelty.[18] Engines in boats served multiple purposes, providing propulsion (increasing their power to trawl heavier nets) and facilitating the use of steel cables and mechanical winches to haul nets (diminishing the manual effort required of fisherfolk and increasing the weight carried by the nets), and the spread of steel-hulled ships led to side trawling being adopted by a variety of fisheries, including cod.[19] On the American coast, trawlers were introduced circa 1905, not without skepticism vis-à-vis the

196

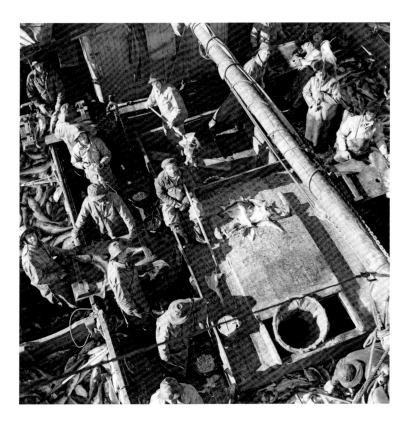

relationship between investment and profit and concerns about their destructive power.[20] When trawling nets drag along the sea-floor, their victims are not merely the fish that they catch: the entire marine habitat is destroyed—including the crustacea and other species on which benthic fish feed—as well as the geological support on which underwater flora and fauna are anchored. Due to their technical complexity, trawlers implied a much larger initial investment, while the fuel they required resulted in higher operational costs. It thus took longer for investments to be amortized and different knowledge was needed on board. Fisherfolk were joined by mechanics. Yet fisheries were faster, landings more voluminous, and the operations profitable. In the cod fisheries, it was a much safer activity than its antecedent, since fisherfolk remained on board a larger ship instead of departing alone in small dories. In the Americas, safety was a key argument in the debate between those who called for trawling to be prohibited and those who stood for trawling as a form of technological and social progress in fisheries.

Cod processing on board lugger *Gazela*, 1953. Photo: Eduardo Lopes. Courtesy Centro Português de Fotografia, EL/247.

Dressing cod on board lugger *Gazela*, 1953. Photo: Eduardo Lopes. Courtesy Centro Português de Fotografia, EL/237.

Between 1909 and 1910, a Lisbon company sent the steam-powered trawler *Elite* to fish the banks off Nova Scotia.[21] This coincided with the adoption of trawlers for long-distance fisheries that had become standard in Atlantic fisheries from the early twentieth century, a technological transformation that reshaped Atlantic fleets after the Great War. Nonetheless, in 1946, the Portuguese long-distance fleet was composed of forty-three ships, twenty-six of them timber-built luggers (twenty with combined propulsion involving sail plus engine), thirteen steel-hulled luggers (all with combined propulsion), and four trawlers.[22] The predominance of the luggers contrasts with the efficiency and capacity of the trawlers, which had a much larger gross tonnage and cargo hold—an indication of the fleet's anachronistic character.

There is a key political event that explains Portugal's perseverance with sailing fisheries. After a military coup in 1926, in 1933 the dictatorship approved a new constitution, the Estado Novo, much inspired by the corporate structure of Italian fascism. In 1934, a regulatory trade commission was instituted for cured cod (Comissão Reguladora do Comércio do Bacalhau, CRCB), a national board securing the economic coordination of the corporate structure and bridging the various interests at play between trade, the processing industry, shipowners, and fisherfolk.[23] The regulatory agency began its activity at a time when a nationalist campaign was seeking to assert an ideological agenda, creating new symbols and unifying narratives. Within this framework, the economic strategy blended with nationalistic mottos, and fisherfolk were paraded as the Atlantic heroes of a timeless tradition of sailing, following a glorious national destiny.

Portuguese fisher with cod, ca. 1940. Courtesy Centro de Documentação de Ílhavo, Museu Marítimo de Ílhavo, Imagoteca, 6431.

The dictatorship smoothed public disagreements with violence, yet beneath the silence there were hesitations and disputes, and understanding the boats allows us to see the extent to which architecture was key to defining the work of the fisheries. In July 1935, the lugger *Santa Joana* was shipwrecked in Greenland. To replace it, the owner commissioned a trawler from a Danish shipyard.[24] The state agency regulating the cod fisheries did not sanction the acquisition and stated that it was "inappropriate" for shipowners to "make any agreements to buy, in foreign shipyards, ships for cod fishing, since no boat can be imported without having a prior visa and permit."[25] If protecting national shipyards was one of the reasons for preventing the import of foreign trawlers, other fears were unemployment and the education required to maneuver larger, mechanized boats. As stated in a subsequent memo, the government acknowledged that "from an economic perspective the advantages of trawling are indisputable," but "from a social perspective it would cause the unemployment of many fisherfolk," and since there were enough fisherfolk for the existing luggers, there was no need to invest in trawlers.[26]

However, not every fisher was happy to board ship for a six-month campaign in the distant waters of the Grand Banks. In April 1937, as a consequence of another legislative bill regulating the cod fisheries and trade, fishers went on strike. The state showed its repressive nature with a decree that mobilized "every fisher that was inscribed in the 1936 campaign," determining that those "who do not present themselves by the deadline announced by the boat captain ... will be punished like [wartime] deserters."[27] It was a hot summer, with a failed bomb attempt against Salazar, the head of state, marking the political instability in parallel to the neighboring Spanish Civil War. It was in this context that the CRCB convened its board in July to assess the "conditions in which the cod fished by the motorized ship *Santa Joana* should be cured."[28] Instead of coming back at the end of August, or even September, the usual time that luggers sailed back from the Grand Banks, the Danish-made trawler requested permission to unload an initial cargo of 240 tons of cod in July, to be processed by Companhia de Pesca Transatlântica, a facility based in Porto. If allowed, the ship would be able to do a second journey to Newfoundland, reinforcing the idea of higher productivity for trawlers as opposed to luggers. The government response relates to the fact that the Portuguese summer was not suitable for drying the salted cod, and beyond the disputes that simmered beneath the surface, the argument demonstrates the key role architecture played in establishing the connection between fishing technology (trawling) and the architecture of processing (the drying racks):

> Given that the season we are now in is especially warm, we decided not to allow the fish to be delivered to the dryers at Companhia de Pesca Transatlântica, Lda.: not only is its location considered dangerous for processing cod during summer, but the results lately obtained in drying imported cod demonstrate a lack of technical competence on the part of its directors.[29]

overleaf Drying racks and processing company Testa & Cunha, Gafanha da Nazaré, ca. 1930. Courtesy Centro de Documentação de Ílhavo, Museu Marítimo de Ílhavo, Imagoteca, 6473.

The story has many underlying layers of complexity: shipowners possessed their own processing facilities, and often, as was the case with Companhia de Pesca Transatlântica, individual entities processed extra cargos for companies that could not cope with the sudden influx of extra fish. In this context, it was the drying racks and the terrestrial equipment that were the subject of technical and political debates. Once the CRCB had taken stock of the *Santa Joana*'s operations, the trade agency found an architectural reason to counter further investment in trawling:

> Some shipowners, far from matching the protection the state is granting to their industry, have, on the contrary, allowed the curing of their cod to deteriorate and are making no apparent effort to improve their facilities or improve their primitive methods. ... To the lack of professional brio on the part of certain shipowners, we should add the poor locations and the lack of capacity of certain dryers to accommodate their fleets.[30]

Drying racks were certainly not central to the ongoing debate, but the architecture built to process cod on land and the way this processing was carried out were key factors in equating the relationship between fishing outlay and investment. Before considering the renovation of the Portuguese fleet, the CRCB organized a commission to "inspect all the dryers and propose suitable measures to allow them to be properly used."[31] The commission's findings expose the political contradictions of marine resource extraction and the terrestrial landscape.

Cod architecture is measured in the relationship between square meters and quintals, a unit used in fisheries for measuring weight, equivalent to 60 kilos according to Portuguese standards.[32] Yet between ecosystems and processing, cod quintals had various weights and measures. After being caught in a line or a net, the fish was beheaded and dressed on the deck by the agile hands of splitters, throaters, and liverers, then salted in the hold, where it remained during the campaign until the ship sailed back to Portugal. Once back, cod was landed as a "green fish," to be later washed and dried before it was ready to be traded. While changing its own physical form, the fish would lose weight and size in every step of the process. In a June 1946 report, a CRCB inspector concluded that "relative to the usual number of workers, green

fish, and drying positions," the area of drying racks required to process cod efficiently was "1 square meter to 1.5 quintals of green fish."[33] We know that to obtain the weight of living fish, green fish weight should be multiplied by three.[34] Each 90 kilos of cod spread out on the drying racks corresponded to 270 kilos of living cod in the ocean. For the average green cod pile in the warehouse, the same report estimates that for each 2 square meter surface, 20 quintals would equate to a height of 65 centimeters.[35] Hence, a 2-meter-high pile occupying a 2-square-meter surface would correspond to 10.8 tons of living cod.

Drying racks gave rise to a specific landscape throughout Portugal, from its northern border with Spain to the southern bank of the Tagus in Lisbon. They were simple constructions, usually slatted structures over granite poles and wooden joists, later replaced by spelter cables stretched from precast-concrete structures.[36] Rows of drying tables were offset, granting access to small wheeled carts and workers, creating large surfaces where cod was laid out in the morning, retrieved by midday when the sun became stronger, put out again during the afternoon, and finally retrieved at the end of the day. The process would last between three and four weeks, according to the weather. While drying resulted from wind, too

Washing the cod before drying, Gafanha da Nazaré. Courtesy Centro de Documentação de Ílhavo, Museu Marítimo de Ílhavo, Imagoteca, 7349.

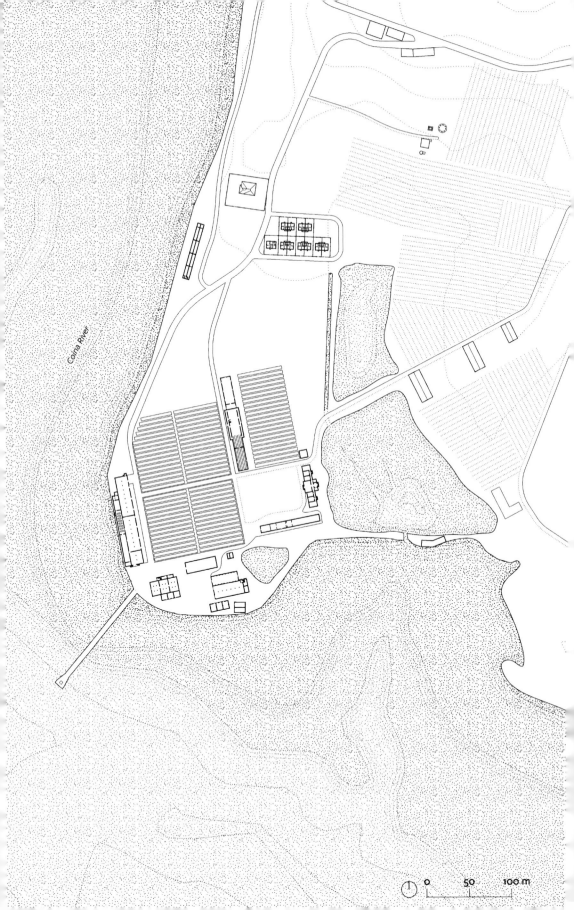

Coina River

0   50   100 m

much sun would burn the cod, and rain would rot it. Hence, the army of temporary workers, almost all women, recruited yearly between September and February, after the fishing season and before the spring sun.[37] In 1957, twenty-nine licensed companies accounted for a total area of 370,000 square meters of drying capacity, with a total of 77,000 quintals of estimated annual production.[38] The average production suggests that we can imagine 41,000 tons of living cod, extracted 5,000 kilometers away, being transformed into 65-centimeter-high piles of cured cod that would occupy a surface area of 1,925 square meters.

Despite the monumental scale of processed fish, facilities were rather simple, providing the stage for a sequence of standard procedures. A primary architectural concern was accessibility: this relied on the existence, or not, of landing docks and piers where green cod could be unloaded from the fishing ships. A key feature here was the warehouses used to store green fish, divided into refrigerated and non-refrigerated locations.[39] The washing tanks were another important element: they constituted a key part of the operation and required a significant amount of water to move the green fish into the dryers. Drying was the central element of the operation. Despite the existence of a few artificial mechanized facilities, natural dryers reigned supreme:[40] the seasonally active fields of drying racks, populated by women, were a distinctive feature of the cod landscape. The drying required more warehouses, to collect, classify, and move commodities. All these elements were complemented by support warehouses and workshops, as well as buildings for paternalistic welfare, such as chapels, kindergartens, and medical offices. These constructions were repeated, with minor variations, in almost all the larger processing facilities.

The example of Parceria Geral de Pescarias (PGP) in Barreiro, on the southern bank of the Tagus facing Lisbon, affords us a detailed look at these patterns.[41] The PGP is located on an elevated platform, not much higher than the high tide, detaching itself from the surrounding hills and providing a large surface exposed to the wind. Having commenced activities in 1873, the company built an ensemble of constructions that, without following a plan or any organized geometry, demarcate the southern limits of the parcel where the pier was located.[42] These constructions combined multiple functions and were built, demolished, and then rebuilt over time, while their functions were adjusted as required. At the western edge of the compound stands a 10- by-100-meter warehouse

facing Cod processing facilities, Parceria Geral de Pescarias, Barreiro, Lisbon, ca. 1950s. Drawing by Gabriel Weber and Daniel Duarte Pereira for Fishing Architecture, 2022.

Building drying racks, SNAB, Cabedelo, Vila Nova de Gaia, 1948. Photo: Cine-Alegre. Courtesy Arquivo Municipal Sophia de Mello Breyner, Vila Nova de Gaia, APRIV/FCA/740/005 (ID 59554).

that, unlike the first group of buildings, incorporates production logic into its design: the green fish would enter through its southern extremity, be stored, and then be washed in tanks outside under the roofs on the sides of the building. Later, an artificial dryer was installed on the first floor. The oldest survey drawings show dormitories for temporary workers on the north side that were relatively detached from the working areas. In the 1930s, a small neighborhood of semidetached houses was built to house the clerks and drivers, with the "boss's house" located in a prominent position. In the 1940s, a monumental, symmetrical building was erected to function as a large artificial dryer (although it was seldom used for this purpose). Canteens and other social amenities were also built in the east of the complex. All these elements gravitated around a central square measuring 160 by 120 meters, a large area of dryers that dominated the complex.

These processing facilities were concentrated in five harbor authorities: Viana do Castelo, Porto, Aveiro, Figueira da Foz, and Lisbon. Lisbon and Aveiro accounted for more than half of production, each being responsible for 33 percent of the total. Porto lagged behind with 20 percent, Figueira da Foz with 9 percent, and Viana do Castelo with 5 percent of total production capacity. Dryers were dispersed in several locations, many of them in quite unexpected places: in Lisbon they could be found in Alcochete, Seixal, and Barreiro, while in Porto the largest facility of all was located on a promontory by the river sandbar.[43] The dispersed nature of these facilities meant that Gafanha da Nazaré, located in Ílhavo's municipality under the jurisdiction of Aveiro harbor, became a reference point within the national context.[44] Gafanha da Nazaré was a desolate location, with loose agricultural production on poor soil with a landscape of sand dunes and salt ponds. The lagoon created by the Vouga River had a very unstable connection to the sea: its mouth was not fixed until the late nineteenth century when a canal was opened to secure safer navigation through the sandbar. The navigation improvements strengthened the harbor service, allowing shipowners to benefit from the development of an important shipyard.

Technically straightforward, cod dryers were simple to assemble and disassemble for seasonal operations, and at the beginning of the twentieth century they spread informally in adapted facilities. Gafanha da Nazaré was in a privileged position to operate long-distance fisheries for several reasons: its safe harbor (despite

its shallow waters) was complemented by a strong local marine fishing culture (centered on the *arte xávega* sardine fisheries) capable of supplying fishing labor with knowledge to supplement the nautical competence of captains and shipowners.[45] In a survey of the 1902 cod fisheries fleet, there were ten ships registered in Lisbon, five in Figueira da Foz, and none in Aveiro.[46] In 1957, there were seventy-two registered boats: eighteen in Lisbon, six in Figueira da Foz, and twenty-seven in Aveiro.[47] While in 1925 there were only two drying facilities in Gafanha da Nazaré, the number quickly increased.[48] Regardless of the reasons leading to the settlement and growth of shipowners and processing facilities in Gafanha da Nazaré during the first half of the twentieth century, the location ended up as a focus for the symbolic dimension of Portugal's cod fisheries.

The 1946 report on the Gafanha da Nazaré dryers contrasted these facilities with the splendid ceremony of the fleet being blessed in Lisbon. Its author, inspector Acácio Castel Branco, commented that he did not understand how "the shipowners [had] not given serious thought to their facilities, leaving some of them looking uncared for or even abandoned in a way that did nothing to recommend them."[49] He even said that some of the facilities should be "absolutely condemned, and some warehouses should

Drying racks of processing company Lusitânia, Figueira da Foz. Courtesy Arquivo Nacional Torre do Tombo, EPJS/SF/005/000368.

be demolished because they were nothing more than shacks in a miserable condition."[50] One of the main reasons for this state of neglect was said to be the timber construction, but another significant factor was the absence of a sewage system. Castelo Branco recommended the construction of drains leading to the river: "This will prevent waste and detritus of various kinds being left to dry, which is unhygienic and a nuisance for others." In the face of the evidence, Castelo Branco confessed to a feeling of impotence: "How can we imagine trying to improve this situation? Only by pulling down all the existing facilities!"[51]

The Gafanha da Nazaré harbor front was carpeted with dryers, with other facilities located in the nearby dunes. Such facilities are visible in aerial photographs showing the intricate relationship between the agricultural fields and the drying racks. The complementary nature of the two labor-intensive activities was obvious, amplified by their seasonality. The CRCB archives show how dryers, especially the most fragile, were built and dismantled with relative frequency, changing owners in constant mergers and transfers. These small facilities occupied the empty spaces in between the major players, where the rational organization of drying racks was more evident. The 1946 report highlights the precarious and muddled conditions that existed between the agricultural areas and the fish-processing industry, recommending that "in the drying areas the planting of any products (corn, potatoes, etc.) should not be allowed during the periods between the departure and arrival of ships, so that the work does not cause the road for cart traffic to fall into disrepair or become covered with mud when it rains."[52] The report explicitly points to the lack of planning as the source of this precarity, but the author blames the conflicts between national, regional, and local administrations pulling in different directions.[53] In 1947, there was a formal urban plan to develop the harbor, reinforce road accessibility, and expand the existing dryers northward. Although the design was not implemented as planned, it nonetheless guided the harbor's development in the following decades, even if many of the fragile wood buildings from the 1920s were still in operation in 1957.

This landscape of cod expanded within the city. In July 1936, when Tenreiro was appointed to oversee the various shipowners' association, from cod to sardines (including trawling), tuna, and other species, the political representation of the Portuguese fisheries set up office on the third floor of 24 Cais do Sodré in Lisbon.

The anonymous building became the center of an intricate network of power linking multiple shipowners' business, trade, and funding policies. Tenreiro stayed in office until 1973, but when he was appointed, CRCB policies were already defined within the framework of the corporate structure conceived by powerful figures of the military regime such as Pedro Teotónio Pereira (1902–1972), Sebastião Ramires (1898–1952), and Ortins de Bettencourt (1892–1969).[54] Since the CRCB was aiming to manage trade, regulate prices, determine imports, and stabilize the ups and downs of supply and demand, providing equilibrium in consumption and managing distribution networks, the political strategy had an architectural consequence: CRCB cold storages.

The technical requirements involved in implementing the policy were settled in a study trip during the summer of 1935. A party of four illustrious Portuguese engineers passed by Holland and Denmark to visit shipyards, stopped in Hamburg, and headed to Norway, where in Bergen and Ålesund they planned to "study cold-storage centers for preserving dried fish."[55] Their architectural goals were explicit: "to study, in conjunction with contractors and building companies specialized in refrigeration, the elaboration of projects for constructing facilities in Portugal that would be similar to those they had seen in Iceland and Norway." In Hamburg they met the architect Heinrich John, who showed them an artificial herring dryer, which the visitors were uncertain would be suited to cod. Although they visited several freezers for fresh fish, their interest was focused on facilities for cured fish, and they reached the conclusion that if the fish were kept below $10°C$, bacteria would not thrive, and under such conditions cured cod could "be kept for many years."[56] It is worth noting that despite the Norwegian parliamentary act of 1932 promoting the construction of freezers, which raised awareness of the low returns of cured fish as opposed to the higher product values of fresh and frozen fish, the Portuguese visitors did not mention in their report the key transformation Norwegian fisheries were trying to implement.[57] Their focus was on drying and preservation methods for cured fish, and the refrigeration technology was seen as a support for the existing business, rather than a means to transform existing practices.

The first CRCB cold storage was built between 1937 and 1939, in Porto. It is a monumental structure facing the river, 45 meters long and 21 meters high. The architect was Manuel Joaquim Norte Júnior (1878–1962), but the entire process was overseen by the

engineer Fernando Yglesias d'Oliveira. The administrative offices were located in a separate building whose shape emphasized the cold storage volume. In the preliminary drawings, the political monumentality was underlined by the inscription "Year XII of National Revolution," although in the final version the choice of typography was scaled down, the building was named after the CRCB, and the national coat of arms took prominence on the façade. The building was structured by means of a concrete grid, with 70-centimeter-thick columns and a central corridor serving seven refrigerated chambers on each floor.

The cold storage was a neighbor of the recently inaugurated Massarelos fish market, built on the basis of a municipal initiative and designed by architect Januário Godinho (1910–1990).[58] The fish market is a point of reference in modern architectural history in Portugal, often quoted for its dynamic design and its ability to combine intricate technical demands with the expression of technological developments. The history of its construction is more complicated than is suggested by simply ascribing it to Godinho's talented authorship. In 1922, the architect Leandro de Morais (1883–1958) designed a regular three-story warehouse with a granite peripheral wall, a concrete structure of columns, and wooden floors and roof trusses. After the successful installation of a refrigerator in the municipal slaughterhouse, in 1932 the municipality launched a competition to adapt the former warehouse to produce ice and preserve fish.[59] The project expanded the building northward with an annex hosting a range of refrigeration equipment, and the interior was divided independently from the original structure. In 1934, when the refrigerator was already in operation, the municipal engineer Correia de Araújo (1909–1981) developed a new project for a fish market contiguous to the original building. It was a large hall with a concrete structure and a glass-tiled roof, connected to the river below street level by a tunnel and a newly built concrete pier. Bidding for the contract, OPCA, an ambitious construction company, presented an alternative proposal based on the design of the young Godinho.[60] This involved a continuous façade that embraced Leandro de Morais's original warehouse, integrated Correia de Araújo's hall into the new design, and used a third administrative building to make the whole asymmetrical and visually expressive. It was Godinho's design that gave the unique architectural qualities to the combination of the three structures (warehouse transformed into refrigerator, fish market hall, and administration). The fish market was inaugurated in 1937, and the entire construction completed in 1938, when the new façade unified the whole in a work that became an emblem for the city.

As local engineer Manuel Bacelar explained, the idea of pairing the fish market with a freezer was to "combat the shortage and insufficiency—in terms of both quality and quantity—of fish for consumption in the city."[61] The building offered three operational systems for processing fish. At street level, the focus was on brine immersion freezing. Fish was washed and dressed and then spread out on trays that would be immersed in brine (for between fifteen minutes and two hours, in a 7- by 3- by 1.5-meter tank) at a temperature of –21.2°C. Once retrieved, the trays were washed, covering the fish with a thin layer of ice, after which the fish were stored in wooden boxes that could be kept for up to six months in a 48-square-meter chamber between –10°C and –15°C. Also at street level, an ice factory produced ice chips to supply trawlers directly via a newly built tunnel and pier. The first floor had two cold chambers, comprising approximately 750 square meters, whose insulation was obtained by a cork encasement between 10 and 17 centimeters thick. Kept between –1°C and –5°C, the chambers were designed to keep up to 250 tons of fresh fish for a "maximum of six days." The largest chamber on the upper floor was refrigerated at between 2°C and 6°C and was intended to keep dried fish; it worked "with the best results until 1939," when it was emptied "because cod started to be stored in the CRCB refrigerator."[62] A few years after initiating its operations, the engineer Bacelar was not optimistic about the facility's performance: "One link of the chain was built without attention being paid to the other links— therefore, so long as fisheries are not regulated and consumer sales developed, this link is more or less useless and stands abandoned and isolated."[63] The failure was a political decision, the state overruling the municipal authorities to foster the centralized and regulated trade of cured cod at the expense of the local trawling activity. A few years after being inaugurated—overtaken by the CRCB cold storage facility—the municipal freezer's main activity was supplying ice to the city, not to the fishing industry.[64]

Municipal cold storage and freezing facilities, Massarelos, Porto, ca. 1936. Photo: Alvão. Courtesy Centro Português de Fotografia, ALV/017111.

The construction of the new cold storages for cod was determined in the 1936 edict that reorganized the CRCB. The legislative bill saw the buildings as key to "constituting permanent reserves" and defined their positions in Lisbon and Porto, since the cities were already established import centers.[65] In Lisbon, the project was in the then new expansion of Alcantara docks. The project was also designed by the engineer Yglesias d'Oliveira, this time with the support of architect João Simões (1908–1993).[66] There were curious tribulations defining the building location and doubts about whose governmental branch would support the investment. Although it was directed by an arm of the infrastructure ministry, it was the Minister of Commerce and Industry who pushed ahead with the process. Fisheries were under the aegis of the navy, not the government, although Tenreiro, a naval official, was favored by the government and not the navy. Built in a strategic position between the city and the expanding harbor, the cold storage took on another layer of symbolic power because of the political intrigue behind its construction. The selected location

Municipal cold storage and freezing facilities, Massarelos, Porto, ca. 1936. Photo: Alvão. Courtesy Centro Português de Fotografia, ALV/017112.

was peculiar since Alcantara, as a new commercial and passenger harbor, did not serve the fisheries. Lisbon's fish market was located to the east, near Tenreiro's office, while the fishing harbor was located to the west, in Pedrouços beyond the Belém Tower. The Pedrouços area belonged to the navy, and these sites first underwent major transformation and development later in the 1960s, when a trawling harbor was built to support the frozen fish industry, which still lacked organization.[67]

Despite being larger, Lisbon's cold storage is similar to Porto's: a nine-story volume measuring 28 by 93 meters, complemented by an independent building housing the administration. The main difference is that Lisbon's was divided into cold chambers for fruit and for cod. As the two sides of the building were equal in size, the split reminds us that a cold storage is not necessarily species driven and could also host other commodities. Yet fish was particular and smelly, and the technical systems of the two areas were completely disconnected to guarantee "absolute separation between rooms that are used for cod and for fruits."[68] At the ends of the building there were lifts and stairways complementing the circulation in the emphasized north-south axis, where the dressing rooms and other complementary areas were also located. Both were designed to store 3,000 tons of cured cod, yet the complementarity between fruit and fish made the Lisbon building look twice as large as Porto's.[69] The building was inaugurated in 1942; the design dated back to April 1938, and the tender was closed one year later. The selected contractor was OPCA, the same company that was then completing the construction of Porto's cold storage and had designed and built the unsuccessful, yet remarkable, Porto freezer and fish market.[70]

Using the new freezing technology to foster old practices instead of promoting the "modernization of fisheries" was a political choice. The government was obsessed with replacing imports with national production, a policy that was supported in part by the outbreak of World War II. When international supply chains were disrupted, Portuguese cod fisheries grew, allowing imports to be decreased and giving the Porto and Lisbon cold storages, where national production was safely stored, the symbolic power conveyed by a successful policy of self-sufficiency. Although the self-promotion implicit in the numbers presented by the CRCB is cause for caution, a statistic published in 1947 emphasizes the growth in the production, value, and consumption of cured cod in

Engineer Eduardo Yglesias and architect Manuel Joaquim Norte Júnior, CRCB cold storage, Massarelos, Porto, 1937–1939. Courtesy Museu Municipal de Etnografia e História da Póvoa de Varzim.

line with the main goal of the decade-old institution: to increase national production and decrease food imports.[71] What the statistics also show is that the decrease in imports was the result of a colossal drop in consumption, from around 1,000 tons in 1935 to 500 tons after 1939. With the progressive growth of the cod fishing fleet (from 9,424 tons of gauge in 1934 to 39,021 tons of gauge in 1947), national production increased from 101 tons in 1934 to 369 tons in 1947. During World War II, the volume of imports decreased from 530 tons in 1939 to 306 tons in 1940 and 277 in 1941, although its value was kept stable by inflation, increasing slightly from 82 million escudos in 1939 to 84 million escudos in 1940. The stability of value camouflaged its impact on consumption, with imports decreasing to 85 percent of overall consumption in 1934, 66 percent in 1939, 54 percent in 1941, and 32 percent in 1947. Yet the success in reducing the volume of cod imports was attributed to CRCB policies, and not to World War II. At the time of their completion, the monumental cold storages were the built symbols of the cod campaign's success.[72]

Food safety was a major problem with cured cod. Severe or fatal cases of food poisoning were not uncommon. In June 1952, a canteen run by Porto social services served up "chickpeas or black-eyed peas with lightly boiled shredded cod, dressed with olive oil, vinegar, salt and onion," resulting in more than 900 people complaining of "headaches, vomiting, and bloody diarrhea."[73] In September 1942, in Riba de Ave, "the workers of a textile factory were poisoned by the cod they had eaten in the canteen, with two of them succumbing."[74]

Most of these cases of poisoning had their origin in "red Bacillus," a bacterial adulteration caused by unsterilized salt.[75] Some, meanwhile, derived from "fisherfolk's furunculosis"—transmitted by the workers handling the fish, from hook to drying racks.[76] One century earlier, before the Portuguese ventured into long-distance cod fisheries, *Battered by Cod*, a satirical booklet of 1826, narrated the misfortunes of one poor fellow who, having suffered several gastric distresses, addressed the fish as follows: "Mister Cod, you are a very rotten, very pestilent and very stinky gentleman"; he claimed that "there was no law forcing [him] to eat what once had been a fish; and Your Pestilence has long since ceased to be one, becoming something else with which the nose cannot be friends."[77] The desperate fellow begged him: "Mister Cod, here is one of your most unfortunate victims, begging you for mercy and asking that you not finish me off with these continuous scourgings, for you have acquainted me with a fair few diseases, the least of them capable of consigning Hercules to his grave."

The eight-page chapbook joined an army of popular literature like the *Sentence of Proscription, Which against Dom Bacalhao Reached Dona Sardinha*—a dialogue on the nationality of fishes, where sardine refers to "Don Cod, arrived from other regions," which "parades amongst us without showing his original looks," and accusing him of being "the original cause of horrific deaths, transmarine follower of inequity practicing every evil most juicily, perfidious marine race."[78] Cod quality varied according to its cure, and many pamphlets complained about decaying quality: "When it was salted with *salt*, it maintained its reputation and commanded respect, but once it started to be salted with *brine*, it lost its good name and credibility, and turned from robust and strong to being valetudinarian and sickly."[79]

These popular references to cod at the beginning of the nineteenth century not only underlined its low quality and the danger it posed to health but also set out to expose the country's economic and food dependency. In 1825, *The Cod's Farewell This Lent* stated that "the Protestant English laugh, filling their pockets with money, while they fill Catholic bellies with cod," insisting that "it should seem like no exaggeration to say that Porto could provide itself with fish from its coasts, for its days of fasting and abstinence: proof of this is supplied by daily observation, so great is the quality and quantity of fish we see that could be dried or salted if this commercial trade was practiced among the Portuguese."[80] Although religion and Lent are often pointed to as the main reason for the local preference for cured cod, the major reasons were political and commercial. Most of the cod trade in Portugal was associated with British business and the colonial economy of port wine, which explains why consumption per capita was much higher in Porto and the northern region, where the textile industry and wine trade were centered.[81]

Shannon Ryan, who wrote an account of the Newfoundland cod trade in the nineteenth century, shows how, despite variations

·A·PESCA·DO·BACALHAU·

cassem á sua pesca n'essas paragens, ao abrigo d'esse celebre tratado.

As tragicas viagens dos Corte-Reaes, nos primeiros anos do seculo XVI, tiveram como consequencia o descobrimento da *Terra Nova dos Bacalhaus*, como lhes chamam os documentos, em cujos bancos iniciámos desde logo as pescarias. D. Manuel, em 1506, por alvará de 14 de outubro, dirigido a Diogo Brandão, manda que este faça arrecadar para o Real Erario o dizimo do *pescado da Terra Nova*, que entrava pelos portos da provincia de Entre Douro e Minho.

Em 1520, o mesmo rei faz doação ao fidalgo minhoto João Alvares Fagundes d'essas terras, e foi em virtude d'esta doação que, entre aquele ano e o de 1525, se estabeleceu na Terra Nova, com gente de Via-

Nós, portuguezes, manifestámos sempre grande predileção pelo *nosso fiel amigo*: o bacalhau, e, talvez, por via d'ele, iniciámos as amistosas relações diplomaticas com a *ncssa fiel aliada*: a Inglaterra!

Com efeito, no meado do seculo XIV, as cidades de Lisboa e Porto celebraram, com Eduardo III de Inglaterra, o importante tratado de 20 de outubro de 1353, que estabelecia, durante cincoenta anos, o direito reciproco de pesca nas costas de Portugal, da Inglaterra e da Brétanha, que n'essa epoca estava sob o dominio inglez. Ora, sendo o bacalhau uma especie que se encontra em alguns pontos da costa ingleza, é possivel que já n'essa epoca os portuguezes se dedi-

na do Castelo, Aveiro e da Terceira, a celebre colonia do Cabo Bretão.

Desenvolve-se então extraordinariamente a

1. Navegando para o banco da Terra Nova—2. Um *dory*—3. Um bacalhau monstruoso, junto dos tanques de lavagem.

and years of similar imports between the two cities, Porto imported 173,000 quintals in 1817 to Lisbon's 155,000, while in 1885 Porto imported 188,000 quintals and Lisbon 67,000.[82] Such commercial preference was explained by specific commercial agreements: in 1810, cod from the United Kingdom benefited from a 15 percent import tax, while a 30 percent tax was levied on fish of other provenance.[83] Assessing national fisheries in 1892, António Baldaque da Silva (1852–1915) accounted for two Portuguese ships in Figueira da Foz and ten in Lisbon that were responsible for unloading 923 tons of cured cod, out of a total of 21,000 tons. Only 4 percent of the consumption was produced by Portuguese fisheries, while 71 percent was unloaded by British ships and 24 percent by Norwegian.[84] From the 1830s onward, the disproportion between imports and consumption propelled various attempts to create a national industry to extract distant resources. The Azores islands, located halfway between Portugal and the Grand Banks, seemed an effective location in which to establish a processing base. An 1840 account describes a comparatively unsuccessful operation involving ten schooners that landed a large cargo in Faial.[85] It was "a big mistake," and "the little amount of cod that was dried, in a humid climate under a strong sun, arrived more grilled than dried."[86] Nonetheless, the Azores continued to occupy a strategic position in Portuguese fisheries: crews were recruited there for the yearly campaigns, and in 1900 a new attempt was made to develop drying facilities on the islands.[87]

These attempts and failures show how difficult it was to build up a Portuguese cod fishery throughout the nineteenth century. Yet it seemed to be part of a strategy to replace imports and achieve a trade balance. The topic inflamed nationalist debates, and in 1894 Baldaque da Silva published a paper claiming the "reestablishment of Portuguese maritime power," in which he said the ultimate purpose was to persevere in the frustrating long-distance fisheries: "Portugal was a great nation of sailors, and only as a nation of sailors will it succeed in becoming great again."[88] From a theoretical perspective, he demonstrates the economic potential of cod fisheries and how the large per capita consumption of cured fish could help to secure funding to build a new fleet, reestablishing the nation's maritime power to the level of "its most sublime historical moment, a beneficial period for civilization, which occurred during the Discoveries, when it was a nation of educated and daring navigators."[89]

facing Page of magazine on national cod fisheries, *Ilustração Portugueza* 368 (March 10, 1913), p. 306. Courtesy BLX-Hemeroteca Municipal de Lisboa.

overleaf Ship routes for cod fisheries. Drawing by Aitor Ochoa Argany and Daniel Duarte Pereira for *Fishing Architecture*, 2020.

Saint Peter's Bank
Green Bank
Misaine Bank
Canso Bank
Banquereau
Rosemary Bank
Le Have Bank
Sable Island Bank
Sambro Bank
Porpoise Bank
Whale Bank
Jaquet Bank
Outer Bank
False Bank
Great Bank
of Newfoundland

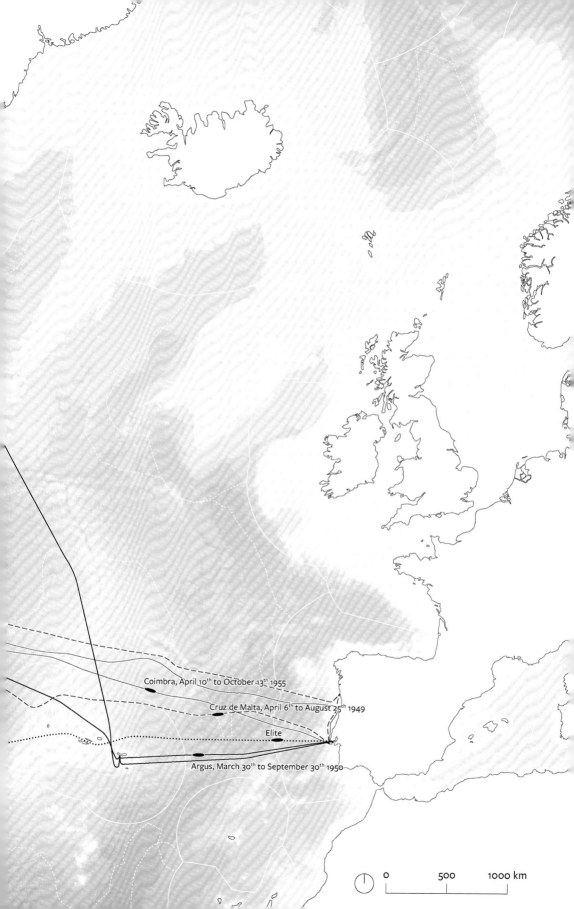

Coimbra, April 10th to October 13th 1955

Cruz de Malta, April 6th to August 25th 1949

Elite

Argus, March 30th to September 30th 1950

0   500   1000 km

Unlike the 1820s and 1830s chapbooks, in which cod was set against sardine, meat, and other animal protein sources, Baldaque da Silva did not aim to replace consumption and suggested taking advantage of the country's eating habits to fund a new long-distance fleet. For such a change, the social position of cod consumption had to change, shifting from a low-cost convenient food to the backbone of a financial strategy to build a new maritime policy.[90] In a 1923 outline of Portugal's cod fishing history, the continuous failures were attributed to fiscal policies that kept encouraging imports.[91] According to economic historian Moses Bensabat Amzalak (1892–1978), the policy was only reversed in 1901, when new customs tax granted protection for Portuguese fishing crews, allowing them to market their produce at competitive prices.[92] In the incongruous variety of statistical charts published for the early twentieth century, with numbers that defy reason and present significant variations of landings and operators, one can nonetheless detect an increase in the number of ships, crews, and landings. In 1917, there were forty ships registered for cod fisheries, a number that decreased to eleven in 1918, only to rise to forty-seven in 1923 before dropping again to seventeen in 1931.[93] Despite national production increasing from 2,300 tons in 1901 to around 5,000 in 1930, the supply was no more than 10 percent of total consumption.[94] In the press, there are plenty of illustrated publications mentioning the "extraordinary progress of this industry in the past few years," a phenomenon always linked with the past successes of sixteenth-century navigation.[95]

It was the Great War, which also brought famine and despair to Portugal, that first blocked British production and limited imports, allowing the country to envision the replacement of cod imports with domestic production. This was the insight granted to a young Salazar, ten years before the military coup that first led to his appointment as minister of finance, enabling him to identify cod—which represented a quarter of food imports—as one of the main reasons for the economic imbalance caused by Portugal's reliance on imported fish.[96] He published a paper that become a reference work, and many of the ideas were addressed again in 1931 in a programmatic report he wrote on Portuguese fisheries.[97] In the second text, he focuses on the sardine-canning industry and again references "some elements in respect to the drying and commerce of cod."[98] Although the 1916 text focuses on the economic problems of the food supply, there is a paragraph that includes cured cod in his political strategy:

> It is a relatively expensive food, seldom accessible to the large proletarian population, who in general have been happy with fish like sardine, when sardine was still a fish for the poor. Nonetheless, to a large percentage of the population, and mainly in the interior regions that rarely, if ever, have a supply of fresh fish, cod is widely consumed and is truly a basic commodity.[99]

In the 1940s, Salazar's argument was gainsaid by reality. The total national consumption was estimated to be 60,000 tons, an annual average of 7.5 kilos per capita.[100] The sales figures for the first six months of 1940 positioned Porto on top, with 5.65 kilos per capita, followed by Lisbon with 3.98, and Braga, Aveiro, and Viana do Castelo all above 3 kilos per capita. At the bottom of the list was Faro, in the Algarve, with 0.15 kilos, accompanied by most of the interior districts where consumption did not even reach 1 kilo per capita. This geography of consumption challenges the idea of cured cod as a staple that was easy to preserve and consumed in regions deprived of fresh fish. In the 1950s and 1960s, this contrast was even more stark, when the national average consumption was 8.8 kilos per capita, but in Porto and Lisbon yearly averages reached 29.6 and 20.8 kilos per capita. Beyond the main cities, other locations with high consumption were industrial centers, such as Matosinhos with 23.8 kilos, Espinho with 19.7, São João da Madeira with 18.4, and Seixal with 15.3.[101] Consumption in other regions, however, was virtually nonexistent. Hence, cod was mostly consumed in urban or industrial locations where other protein sources were available, including fresh fish. This access came at a price: the cured cod, whose supply was stabilized by the large Porto and Lisbon cold storages, offered protein to a population who relied on store-bought food, without access to vegetable gardens, orchards, or domestic poultry, subsistence food supplies that were accessible in non-urbanized areas.

With the construction of Porto's fish market and freezer, inaugurated in 1934 on the eve of the CRCB's incorporation, the municipality attempted to augment the urban food supply with fresh or frozen fish. Yet the state dictatorship favored cured cod as an element of structural subsistence that would also foster a national identity. While cod was targeted to ameliorate the balance of trade, sardine canneries—an industry also supported by the government—aimed to export canned products. Both trawling and freezing, which required a shift both in production systems (different ships and technical knowledge) and in consumption habits (different support infrastructures and promotional campaigns), were downplayed in favor of cod. A major reason for this political decision seems to have been the nationalist narrative that had emerged from the nineteenth-century debates and consolidated in the early twentieth century. Beyond supplying food needs, cod had become a mark of national identity, capable of evoking the mythology of the age of discovery while also serving as a food staple that represented a social structure. The drying racks and cold storage architecture that were made possible by the muscular white flesh of the Grand Banks cod were the physical imprint of this nationalist policy.

The transformation of fishing technology sheds light on a deeper cultural divide, explicit in literature and cinema. In Georges Simenon's 1931 crime novel *Au rendez-vous des Terre-Neuvas*, the famous inspector Maigret investigates a crime on board a French cod trawler in Fécamp.[102] The *Océan* had spent three months fishing in the Grand Banks off Newfoundland, and besides the usual adventures within the novel, Simenon narrates the hardship and violence on board, although this did not compare with the outrageous four-to-six-month campaigns that were still occurring on sailing boats. In 1931, the year in which the novel is set, there were still four sailing ships using dories in Fécamp's cod fisheries.[103] In 1937, *Captains Courageous*, a grand US production directed by Victor Fleming (1889–1949), had its première. The feature was a screen adaptation of Rudyard Kipling's (1865–1936) eponymous literary success, first published in 1897.[104] From the experience of US fisheries in the Grand Banks, Kipling described the hardship of fisherfolk and praised their courage while narrating the formative adventure of a new hero, a rich and spoiled kid who, after falling off a cruise ship, was saved from drowning by a schooner, on which he afterward learned the value of selfless work. Kipling's novel is part of a romantic genre that helped to promote the cult of genius, heroes, and supermen, as well as the presumption of supremacy of a select breed capable of dominating the tense relationship between progress and nature.[105]

French cod trawler *Bois Rosé*, Fyllas Bank, Greenland, 1952.
Photo: Anita Conti. Courtesy Archives Lorient, VU'.

In 1950, following in the spirit of Kipling's novel, Pedro Teotónio Pereira, a major political figure in Portugal, commissioned
Alan Villiers (1903–1982), an Australian maritime journalist, to
produce a report on the Portuguese cod fisheries.[106] The book,
written in the wake of a three-month journey on board the lugger *Argus*, became famous, was translated into several languages,
and was accompanied by the production of a documentary film,
whose visual impact still endures.[107] In the last scenes of his adventures, Villiers quoted the Portuguese nationalist adage: "The simplicity of life, the purity of custom, the gentleness of feeling, the
equilibrium in social relations, so modest but so dignified." While
observing the luggers anchored in Lisbon, he was reassured "by a
fine maritime Nation, rich in the traditions of great voyages, which
refused to be disrupted by the craze for mechanization and the
trend towards barbarism" and assured the reader that in those
ships "were these things all symbolized: here were their fruits in
the shape of good ships, good captains, good men, and an abundant harvest brought in from the sea."[108]

Villiers's epic tale contrasts with a contemporary French editorial success, the book *Racleurs d'océans*, published in 1953 by the
oceanographer Anita Conti (1899–1997). In 1952, Conti had gone on
board a French trawler in its Greenland fisheries.[109] While Villiers
admired the vestiges of a disappearing world, viewing with ten-
derness the maritime toil of the Portuguese sailors, Conti focused
on the characteristics of the marine ecosystem and its relationship
with the work of fishing.[110] In her descriptions of life at sea, Conti
admired the strength and endurance of the fisher—but her focus
was on the relationship between activities on board the boat and
the environment, the geological and ecological qualities of the seafloor.[111] An example of such care is her description of the capture
of a cod shoal on an underwater ridge: "The ships traced endless
threads over lines of water with the same depth, 62 to 65 meters.
The cod is there, probably in a water current whose temperature
suits it and the other creatures it is chasing to feed on."[112] Conti is
aware how trawlers boosted the extractivist dimension of fisheries and emphasizes that, unlike agriculture or industry, fisheries
did not produce but only killed: "A boat spots the fish, captures
and kills it, a fishing boat is a hunter and an industry, never a producer. ... In the oceanic environment we exploit, blindly."[113]

Conti's work echoes Rachel Carson's (1907–1964) *The Sea Around
Us*, a marine biology book that made it onto the *New York Times*

226

bestseller list in 1951.[114] Carson had an extraordinary impact on the dissemination of science and on the public understanding of the sea and its ecosystems, an ecological dimension amplified in 1962 by her *Silent Spring*, a research work on the environmental effects of synthetic pesticides, whose publication is often used as a reference for a new form of ecological awareness.[115] Carson spotted a problematic conflict: "We still talk in terms of conquest. We still haven't become mature enough to think of ourselves as only a very tiny part of a vast and incredible universe."[116] While progress was still seen as an instrument for domination, scientific knowledge was perceived as a tool for democracy capable of overcoming the cleavage between humanity and nature. Carson argued: "The materials of science are the materials of life itself. Science is part of the reality of living; it is the what, the how, and the why of everything in our experience. It is impossible to understand man without understanding his environment and the forces that have molded him physically and mentally."[117] In the aftermath of the technological shock of World War II, the limits and dangers of technology had become visible, as well as the need to reposition science to promote human welfare. Appearing at the same time as Carson's work, Villiers's and Conti's books are not parallel depictions of sailing schooners and trawlers or setlines and trawls; rather, they express two different visions of the relationship between society and nature.

The shift from the lugger or schooner to the trawler had an immediate territorial impact. Setline fishing exerted greater pressure than handlines, to the point that Newfoundland coastal fisheries moved offshore and explored new fishing banks on the high seas that were farther and farther away. Back in the nineteenth century, schooners and dories had already exhausted abundant fishing banks like Banquereau off Nova Scotia and pushed sailing routes farther away.[118] Trawlers accelerated the process. If we compare the maritime routes of Portuguese ships, we have a good picture of the continuity of this process.[119] In 1894, the lugger *Labrador* sailed over the Jacquet Bank and the Grand Bank without mooring in Newfoundland from the end of May until October, when it sailed back to Lisbon, stopping in the Azores.[120] In 1910, the coal steamer *Elite* worked between May 27 and June 25 between Canso Bank, Misaine Bank, and Green Bank off Nova Scotia, where it moored to refuel in North Sydney, Cape Breton Island.[121] The *Elite* had less than a month's worth of fuel autonomy and higher energy costs

Hauling the cod trawl, *Bois Rosé*, Fyllas Bank, Greenland, 1952.
Photo: Anita Conti. Courtesy Archives Lorient, VU'.

as compared with a sailing vessel.[122] But what becomes evident in the navigation routes from the 1930s onward is that trawlers or luggers, after a few weeks harvesting in the Grand Banks off Newfoundland and as soon as weather conditions allowed, would set sail—usually in early July—to Fyllas Bank off Greenland. This new territorialization of fisheries, which could be clearly seen when trawling became a standard practice in the French and American fleets from 1905 onward, was a consequence of the ecological pressure created by technological transformations.

This territorial shift repositioning fisheries geography had a subtle effect on the urban landscape. One of the effects of overfishing is the scarcity of older and larger animals: there is a progressive decrease in the average sizes of fish, which become not only younger but also smaller. A 1966 photography of the Terranova grocery shop in the center of Lisbon shows a variety of cured cod hanging in the storefront.[123] While Icelandic and Norwegian cod cost $30 and $28 per kilo, the "national" Portuguese cod caught in Newfoundland only brought in $26 and was substantially smaller. In the lower piles of fish in the picture, there were even smaller cod, whose price ranged from $18.50 to $20. It is easy to understand that fishing increases the mortality rate of fish, and once the mortality rate is higher than recruitment (the number of fish that evolve from juvenile to adulthood and become sexually mature), a given population declines. A sign of such a decline is the decrease in the size of captured individuals, with instances of adults surviving until they grow large becoming rare.[124] If in the nineteenth century it was normal to see individuals as large as 1.7 meters in length,[125] the dimensions of large fishes progressively decreased, as we can see in the Lisbon shopfront. It is worth recalling that overfishing of this kind in the Grand Banks was not occasioned by Portugal's outdated fisheries but was a consequence of the combined efforts of the American and British fleets, and, from the 1960s, Eastern European countries, amplified by the introduction of onboard freezing and factory ships.[126] What is also evident is the relative stability of cod from Iceland and Norway, whose populations were still less pressured than those of Newfoundland, where the "national" Portuguese cod was then harvested. The decline of the cod architecture built in Portugal during the twentieth century is associated with the decline in the cod population in Newfoundland, propelled by trawling fisheries. While it clearly shows the size difference between the small

Newfoundland cod and their larger Norwegian and Icelandic relatives, the image is a striking portrait of the ongoing overfishing in the Grand Banks and the intimate relationship between marine ecosystems and terrestrial landscapes.

The landscape of Portugal's cod-drying facilities has become a sublime ruin. Finally abandoned in the early 1980s, the areas of prefabricated concrete poles remained on hold on the outskirts of several cities. The racks' spelter cables broke and become rusty. Chased by the wind, vegetation covered the ground in between the racks, warehouses were abandoned, paint decayed, and roofs collapsed. In the past decade, real-estate pressure on riverside and urban plots fronting the ocean finally arrived at these marginal locations, and many sites were cleared, giving way to high-end housing complexes. Yet the memory of these fragile and precarious landscapes where fish was processed is still vivid. The ruin of architecture retains a strange magnetism; nostalgic attempts are made to classify the physical remains as heritage sites, as visible remnants of bygone eras. It is an elusive temptation. In this amphibious history, there is no marine counterpart to the picturesque quality of terrestrial ruins. A major factor in the decay of cod landscapes was the depletion of cod populations, and that is invisible for the most part. Fish is vanishing from the Atlantic. The visibility of architectural ruins corresponds with the invisibility of ruined marine ecosystems.

Follow the fish. Fisheries are responsible for a history of devastating transformations of marine ecosystems, a history of depletion and ecological destruction.[1] Fish tell tales of how cultural and technological changes brought about both human wealth and human misery that challenge established logics of economy and progress. Looking at the destruction fisheries have left in their wake from an architectural perspective, it becomes obvious that the built environment of fishing provides significant evidence of ecological conflict between an extractive industry and the physiological and environmental conditions of marine species. In a history of fishing architecture that traces changes over time, we can see how construction results from the biological traits of marine species and reflects the status of oceanic ecosystems, regardless of whether it is considered vernacular or modern. Following the fish, settlement patterns can be understood according to the configuration of underwater ridges, water temperature, and relative abundance of species, while the decay of certain building forms can be seen as a consequence of overexploitation and exhaustion of natural resources. Biology, technology, processing, politics, and consumption are lenses to help us perceive just how intimately entangled fish and terrestrial landscapes are.

This entanglement calls for an amphibious history that relates to both land and water, and the primary goal of this book is to do just that: by drawing on work done in the growing field of socioecology, it shows how ecological factors motivate building practices and studies the impact of architecture upon the natural resources of the North Atlantic. In the long term, the hope is to establish this amphibious approach within architectural history. Among its precedents is Sigfried Giedion's chapter on slaughterhouses in *Mechanization Takes Command*. He brought the anonymous tradition to the forefront of architectural debate, describing the slaughterhouse as a bridge between animal husbandry and the assembly line that delivered cows and hogs into the mechanical purview of modern architecture, while introducing them into the glossy pages of architectural discourse.[2] *Mechanization Takes Command* describes the building processes of the assembly line and of serial production and discusses the exploitation and mechanization of nature, putting meat at the center of the architectural narrative. The slaughterhouse as the locus for the conversion of an animal into a commodity is examined critically by William Cronon in *Nature's Metropolis*, where he relates the world-famous Chicago slaughterhouses to the American landscapes that were the source of the meat they processed.[3] We read of such things as the extinction of bison in Nebraska and Wyoming and cowboys driving cattle north from Texas to Chicago. As in fishing architecture, the slaughterhouse and the city relate to various species and to large territories across which ecosystems are linked to urban markets. Whereas Giedion described a connection between the slaughterhouse and design practice, Cronon demonstrated its impact on the larger environment in a chapter with the ominous title "Annihilating Space."

In these accounts, land occupies center stage. Today, scholars are building an environmental history of architecture, tracing the sources of construction materials around the globe.[4] In these accounts, oceans are either connecting spaces across which materials move or a locus for the extraction of valuable natural resources like oil and minerals. The groundbreaking research by Nancy Couling and Carola Hein makes the ocean an architectural topic by looking at how it is territorialized by trade routes, mineral extraction, and energy production.[5] And, as in other recent cross-disciplinary research work on the oceans, we see arts disciplines leading the way in linking science and culture.[6] Still, despite an awareness of the ecological impact of human actions, the links between marine life and urbanization are yet to be tackled or, in some cases, noticed. From an architectural perspective, what happens on land stays on land, and what happens in the ocean stays in the ocean. The assumption is that buildings can only be understood in the context of their direct vicinity, assessed from a human or sociological angle and often regardless of how humans relate to other species. In that regard, fish cannot be further away from building practices, yet much architecture has been built to follow fish. And it has certainly had an impact on fish populations.

Considering the hypothesis of an amphibious architecture starts with fish and the marine ecosystems in which they live. The ocean has a variety of environmental factors in constant mutation, among them water salinity and temperature, currents, and the geological character of the seabed. These are accompanied by equally complex ecological conditions. The relationships between species in food chains, ranging from plankton to the higher trophic

234

levels, vary according to predatorial behavior and seasonal dynamics. And as for the fish species itself, variations in individual and social behavior, dietary preferences, spawning habits, growth rate, migration patterns, population distribution, and genetic codes all play important roles. Understanding such dynamics makes it possible to correlate architectural events with ecological episodes.

As a human activity, fishing is better documented than marine ecosystems and can be inscribed within the history of technology. Fishing history combines the histories of navigation, fishing gear, and many other innovations that affected how many fish were caught. Thus it is a useful entry point through which to begin looking at the sea-land interface. For example, how boat docks set infrastructural requirements can be a key piece of information for scholars studying the establishment of settlements, the impact of artificial harbors, and the physical relationship between fishing grounds and workforce. The scale of the vessel and the fishing effort it implies relate to capital needs and energy consumption, to the scale of the workforce, and to the related fishing output. Fishing vessels were transformed historically by the introduction of steam power, steel hulls, and diesel motors, technologies that had a major effect on catches. These changed the time and energy required to travel from the harbor of origin to the fishing grounds and back, crucial factors in the relationship between architectural output and ecological pressure. The more efficient the fishing gear and available technologies, from navigation systems and tools, fishing lines, and nets to the maps, charts, and sonar devices used to locate the prey, the more intense the pressure the fishery exerts on the ecosystem. The fishing gear used also defines the crucial moment of capture, when the balance of environment and ecosystem is actively interfered with, as well as the results of this interference. And although this equipment might seem far from architecture, larger and faster fish landings require the building of higher-capacity processing facilities.

In any fishery, food processing is the pivot between natural resource and consumer product, and so processing facilities like drying racks and canneries are the most obvious architectural by-product of fishing. An amphibious architectural history encompasses the production chain, a network of systems that starts in the boat when the fish is caught and ends when the edible merchandise is shipped to consumers. Studying the nature of each link in this chain and its repetitive mechanics provides an architectural framework through which we can assess the implications of the fishery. For instance, the introduction of freezers brought the most dramatic historical shift yet in the shape of fishing architecture, with an impact on both landscape and fish populations. This is not just because the monumental presence of cold storage and freezing plants in ports and fishing villages changed landscapes previously defined by drying racks, but also because the use of freezers at various stages in the production and consumption chains led to unprecedented increases in the distance between the ecosystem where a fish lives and where it is consumed, with a single ecosystem serving an ever-growing population. We know of the ecological impact of such transformation; its architectural implications are part of the amphibious history tackled in this book.

An environmental history of architecture cannot be constructed within conventional national boundaries because despite the national—if not nationalistic—context of industrial investments, fish have no nationality. Nonetheless, most of the data and architectural phenomena assessed in this book are still explained with reference to specific geographic and sociopolitical contexts. Politics inserts itself between fish and fishing architecture and interferes with the ecological pressure exerted upon fish populations. Hence, an amphibious history of Atlantic architecture that transcends local biases requires us to confront Atlantic fishing politics. This involves territorial disputes over maritime sovereignty and fishing rights; state support of fishing policies and technologies, targeting specific species; the ties between the fishing industry and war efforts; national, cultural, and economic policies that foster biased narratives; and folklore and heritage that nurture a specific cultural image of fishing. Most significant is the insignificance of terrestrial politics to the objective characteristics of fish populations and the health of marine ecosystems. The shifting lines of maritime sovereignty are indicative of major fishing disputes, with various countries claiming their right to access fishing grounds. Such disputes allow us to follow the fish to shore and to answer the question of how the paths of two fish from the same shoal, captured by vessels of different nationalities, will differ both geographically and architecturally. Yet the fish do not care about political boundaries or identity disputes.

And, finally, we eat fish. Whether fresh, dried, salted, canned, or frozen, fish is consumed in many ways, and how we eat affects ecosystems and their architectural counterparts. Consumption habits dictate the urban expression of fisheries—from market stalls and fishmongers to supermarket freezers—and put demands on processing chains. And consumer behavior itself is shaped by legislation, investment policies, advertising, and the cost of fish. The latter is a key factor in determining demand because the price consumers are willing to pay at the market affects how much fish is needed and processed. And the fluctuating relationship between consumer demand and the available supply determines the market value of fish, placing an economic value on an ecological system. This relationship seems to have no architectural manifestation, but it is within these economic dynamics that architecture operates, and thus practices that shape consumer behavior also affect fishing architecture. For example, increasing demand propels industrial development and favors construction. Once built, factories and processing facilities have a powerful inertia and require increasing consumption to secure their economic survival and growth.

By layering five analytical perspectives—biology, technology, processing, politics, and consumption—environmental histories of the ocean can be extended into the realm of architecture and lead to further understanding of the unnatural history of the seas. An amphibious gaze transcends histories of architecture as an exclusively human activity or as an art form, instead seeing it as a component within a larger socioecological history.

For centuries, the dominant mode of architectural representation has followed the relatively simple convention of defining a project by a set of scaled plans, sections, and elevations. Vernacular innovations, prompted by unconventional empirical knowledge, have always challenged the norm, but it wasn't until the introduction of computer software that links building design directly to production systems that we see a radical inversion in the process.[7] Digital technologies brought architecture and industry closer, enabling a parametric architecture in which forms are generated by data and variables in constant mutation. Despite inertia and many difficulties, this eventually expanded the discipline along with distancing it from classical systems of representation that assume a static unity between form and time.

Architects know that the life of a building starts long before they get involved, when the client figures out a brief or considers the geomorphology of a site to be built on. All architecture is ephemeral, and yet a building's life never really ends. A building is in its prime when it is in use, after the work of the architect is done and construction is complete. Once built, however, it is subject to change. It might be added to, moved, or demolished and its stones reused in other buildings. Nonetheless, despite the uncertainty of a building's future, the time of its design is relatively stable: a building is conceived to be built at a specific moment in a specific place. Over the last several centuries, it is this design project—codified in its representation and stable in space and time—that has defined architecture.

A fish shoal is, by definition, an unstable entity. A fish life cycle implies change as the fish develop, migrate throughout the oceans, and occupy different spaces according to the seasons of the year and even the time of day. Unlike architecture, usually built with inert materials, fish are in constant movement through time and space. So how can we represent the ecological dynamics of fish alongside the physical transformations of architecture? How can we reconcile a discipline trained to create stability with unstable life in permanent mutation?

The nineteenth century equated progress with built works and public facilities that improved people's living conditions and with fantastic infrastructure that conquered natural barriers to make remote places accessible. But in the twenty-first century, we see the obvious frailty of so-called progress that relies on the exploitation of natural resources. Gains in the quality of life of human populations put an unbearable environmental load on other species and on the planet's ecological systems. Reinventing architecture by reconfiguring the parameters of its relation to the world and to its own history is now imperative. We need a history capable of unveiling the dynamics and interdependencies between what is built and its ecological impact to support the development of a more conscious architectural practice that is in tune with nature. And although it might seem awkward to seek to understand how the biological characteristics of fish have affected and defined landscapes, such an approach is crucial in defining new modes of architectural thinking for the future. Thus, it is worth trying to synchronize architecture with fish dynamics as an open hypothesis of a new approach.

Herring oil factory, Øksfjord, Norway, ca. 1935.
Courtesy Finnmark Fylkesbibliotek, FBib.00020-014.

overleaf Canning factories chimneys, Setúbal, Portugal, 1929. Photo: Américo Ribeiro.
Courtesy Câmara Municipal de Setúbal, AFAMR/AMR-14561.

Acknowledgments

This book may have originated with my childhood memories of decaying cod drying racks over the sand dunes in São Jacinto. While the research process started in 2017, my first acknowledgment goes back to 2015 and the anonymous jury of a long-term postdoctoral fellowship with Fundação para a Ciência and Tecnologia (FCT), sponsored by the Portuguese Ministry of Science and Education. At a time when most people would laugh at my hypotheses, their funding provided the conditions in which I could explore an unknown field. Hosted in the School of Architecture of Minho University in its Lab2PT, Landscape, Heritage, and Territory Laboratory, I had the privilege of starting to build a research team with dedicated colleagues who, in various stages of the process, contributed to the research and expanded its scope and ambition. Diego Inglez de Souza was a most congenial companion on the Portuguese research front, and this book draws on a good deal of the conceptual framework devised for the book *Arquitectura do bacalhau e outras espécies*, which we coauthored and published in 2022. Aitor Ochoa Argany was an early member of the team, and Daniel Duarte Pereira expanded its interests to sargasso and sea algae, while José Pedro Fernandes focused on a case study relating to contemporary fisheries. We were supported by a group of biologists from CIIMAR (Interdisciplinary Centre of Marine and Environmental Research) at the University of Porto, and we are especially grateful for the advice of Elsa Froufe, Filipe Castro, and Francisco Arenas, as well as Mónica Felício and Diana Feijó from IPMA (Portuguese Institute for Sea and Atmosphere).

In 2020, we opened the exhibition *Our Land Is the Sea* at Garagem Sul in Centro Cultural de Belém (CCB), just a few days before lockdown. The long-term collaboration I had with this Lisbon cultural venue was fundamental in nurturing these, and other, research topics. I am grateful to Isabel Cordeiro and Madalena Reis for the freedom they grant me to explore architecture as a form of knowledge, and to the wonderful team where Margarida Ventosa, Diogo Nunes, and Inês Marques excelled in engagement to achieve high standards with restricted means. The exhibition on the sea I co-curated with Miguel Figueira was a collective experiment on the Portuguese shore. Many ideas moved from the exhibition to this book, and a large thank you is due to Pedro Maurício Borges, Marta Labastida, José Albergaria, Rik Bas Backer, Ivo Poças Martins, Pedro Bandeira, Eurico Gonçalves, and Carla Cardoso. Following the exhibition, I benefited from the support of Béatrice Leanza and the MAAT (Museum of Art, Architecture and Technology), which allowed me to stage a productive workshop, while the EEA's Fund for Bilateral Relations enabled a fruitful research trip to be undertaken in Norway.

At an early stage of the research, Sébastien Marot recognized its potential and fostered an exchange program at the Observatoire de la condition suburbaine (OCS), a research center operated by Université Paris-Est. At the OCS, I was welcomed as a fellow funded by the Project I-Site of Gustave Eiffel University, and I benefited from a lively research environment and pertinent discussions: I would also like to express my particular gratitude to Paul Landauer, Luc Baboulet, Éric Alonzo, and Frédérique Mocquet.

There were many people who provided guidance in the context of the different research trips and fieldwork, reading manuscripts, suggesting ideas, and putting forward hypotheses at various moments of the research. We learned from fisherfolk, librarians, scholars, and friends, and it would be impossible to name them all. Yet some contributions proved key to completing this book: Karl Otto Ellefsen, Mari Hvattum, and Navid Navid in Norway; Paul Dean, Sheila Devine, and Robert Mellin in Newfoundland; and Marta Macedo and Inês Amorim in Portugal; as well as Jennifer Hubbard, Callum Roberts, George Kapelos, Daniel D. Barber, and Loren McClenachan in many emails. João Faria is a longtime ally in the shaping of my books, and, as always, his support and advice proved crucial in handling the manuscript's visual dimensions.

While the book was in the making, I had the opportunity to be hosted by the research center of the Faculty of Architecture of the University of Porto, followed by a generous consolidator grant from the European Research Council. The new research context was a boon to push the work further, and I am thankful to João Pedro Xavier, Pedro Rodrigues, and José Miguel Rodrigues for the university's unconditional support of my research. A new research group is being set up, and the engagement and support of Alice Nouvet, Rafael Sousa Santos, Paul Montgomery, Ana Costa, Claudia Soares, Sónia Gabriel and Garðar Eyjólfsson were key in the last phases of the writing, as were the contributions of Alex Jordan, Nancy Couling, Matthew Gollock, and Nuno Queiroz.

To overcome my clumsy English writing, I have been lucky to have encountered generous readers, and this book would have not existed without the engagement, care, and patience of Megan Spriggs and Simon Cowper. Their dedication to my rough manuscripts inspires me to keep on writing and bring conflictive ideas into a possible narrative. Last but not least, this book would not have been possible without the confidence and enthusiasm of Thomas Weaver, extending from when the book was nothing more than an idea until it emerged as a printed object.

The publication of this book was made possible by a grant from the Graham Foundation for Advanced Studies in the Fine Arts.

Fish sales at Ribeira Nova market, Lisbon, ca. 1910. Photo: Joshua Benoliel.
Courtesy Arquivo Fotográfico de Lisboa, PCSP/004/JBN000765.

overleaf Fish shop storefront, Lavender Hill, London, ca. 1914.
Courtesy Thislife pictures/Alamy 2HERDRF.

Notes

Prologue

1. Karl Otto Ellefsen and Tarald Lundevall, *North Atlantic Coast: A Monography of Place* (Oslo: Pax Forlag, 2019), 174.

2. On the postwar cod landscape in Norway, see chapter 4, "The Salt and the Freezer." Ellefsen and Lundevall provide a detailed analysis of the modernizing ideals that led to the construction of the Fryseriet, transforming a harbor of marginal importance within the network of northern fishing communities into a case study of the welfare state in postwar Norway.

3. The modern building also represented the dominance of motor-propelled trawling gear, as opposed to the various techniques that had come before, from lines, jigs, and gill nets to seining. Although this dominance introduced new boats and fishing practices, and along with them a new system of relations among fishers, it was an existing technology that inherited much of its knowledge from previous practices.

4. Henri-Louis Duhamel du Monceau and L. H. de la Marre, *Traité général des pesches, et histoire des poissons qu'elles fournissent, tant pour la subsistance des hommes, que pour plusieurs autres usages qui ont rapport aux arts et au commerce*, 3 vols. (Paris: Saillant & Nyon/Desaint, 1769–1782). See vol. 2 (1772), part III, p. 487, and plate xv, and the text on pages 413–414.

5. On the architectural strategies of the encyclopaedia drawings, see André Tavares, *The Anatomy of the Architectural Book* (Zurich: Lars Muller Publishers/Canadian Centre for Architecture, 2016), 292–315, and Roland Barthes, "Les Planches de l'Encyclopédie," in *Le Degré zéro de l'écriture, suivi de Nouveaux essais critiques* (1964; Paris: Éditions du Seuil, 1972), 89–104, translated by Richard Howard as "The Plates of the Encyclopedia," in *New Critical Essays* (New York: Hill and Wang, 1980).

6. For a detailed analysis of a British herring smoking house, see Matthew Bristow, "The Pre-Industrial Lowestoft Fish Office: Reading Socio-Political Events through a Vernacular Building," in *Vernacular Architecture* 51, no. 1 (2020): 1–22.

7. A wealth of films and documentaries portray Mediterranean and Atlantic tuna fisheries—in Italy, *Contadini del mare* (1955, directed by Vittorio de Seta); in Spain, produced by the Consorcio Nacional Almadrabero, *Almadrabas* (1933, directed by Carlos Velo and F. J. Mantilla) and *Costas del Sur* (1956); and in Portugal, *A pesca do atum* (1939, directed by Leitão de Barros), *Almadraba atuneira* (1961, directed by António Campos), and *O Copejo* (1969, directed by Hélder Mendes for national television). With epic undertones, they all illustrate the complex operations involved in managing the tuna traps.

8. António Baldaque da Silva, *Estado actual das pescas em Portugal compreendendo a pesca marítima, fluvial e lacustre em todo o continente do reino, referido ao anno de 1886* (Lisbon: Imprensa Nacional, 1891), 221.

9. Bernard Rudofsky, *Architecture without Architects: A Short Introduction to Non-pedigreed Architecture* (New York: Museum of Modern Art, 1965).

10. Ian Urbina, *The Outlaw Ocean: Crime and Survival in the Last Untamed Frontier* (London: The Bodley Head, 2019).

11. A Portuguese forerunner to this book, written in collaboration with Diego Inglez de Souza, was published in 2022. André Tavares and Diego Inglez de Souza, *Arquitectura do bacalhau e outras espécies: Uma leitura crítica da paisagem construída pelas pescas portuguesas* (Porto: Dafne Editora, 2022). The preceding paragraphs on tuna and sections of the prologue, epilogue and chapter 5 are abridged translations from that earlier work.

12. Jón Jónsson, "Fisheries off Iceland, 1600–1900," ICES *Marine Science Symposia* 198 (1994): 3–16.

13. Alexandra Silva, "Morphometric Variation among Sardine (*Sardina pilchardus*) Populations from the Northeastern Atlantic and the Western Mediterranean," *ICES Journal of Marine Science* 60 (2003): 1352–1360; R. H. Parrish, Rodolfo Serra, and William Stewart Grant, "The Monotypic Sardines, *Sardina* and *Sardinops*: Their Taxonomy, Distribution, Stock Structure, and Zoogeography," *Canadian Journal of Fisheries and Aquatic Sciences* 46 (1989): 2019–2036. See also Mário Ruivo, "Sobre as populações e migrações da

sardinha ('Clupea pilchardus' Walb.) da costa portuguesa," *Boletim da sociedade portuguesa de ciências naturais*, 2nd ser., 3 (1950): 89–122.

14. David J. Starkey, Jón Th. Thór, and Ingo Heidbrink, eds., *A History of the North Atlantic Fisheries*, vol. 1, *From Early Times to the Mid-nineteenth Century* (Bremerhaven: Deutsches Schifffahrtsmuseum, 2009); David J. Starkey and Ingo Heidbrink, eds., *A History of the North Atlantic Fisheries*, vol. 2, *From the 1850s to the Early Twentieth-First Century* (Bremerhaven: Deutsches Schifffahrtsmuseum, 2012).

15. Mark Kurlansky, *Cod: A Biography of the Fish That Changed the World* (New York: Walker, 1997); Poul Holm et al., "The North Atlantic Fish Revolution (ca. AD 1500)," *Quaternary Research* (2019): 1–15.

16. Plankton, which absorbs solar energy, is eaten by fodder fish such as capelin (*Mallotus villosus*), herring, or sardine, which in turn become the food of carnivorous fish such as haddock, cod, and tuna. These fish are now being transformed: into food for other fish species in aquaculture, into fish meal for other animals, and even into agricultural fertilizer (reversing the direction of the trophic chain).

17. Jason S. Link, Bjarte Bogstad, Henrik Sparholt, and George R. Lilly, "Trophic Role of Atlantic Cod in the Ecosystem," *Fish and Fisheries* 10, no. 1 (March 2009): 58–87.

18. Railways boosted the trawling fisheries, helping to increase catches and the consumption of fish during the nineteenth century. On railways and trawling, see Robb Robinson, *Trawling: The Rise and Fall of the British Trawl Fishery* (Exeter: University of Exeter Press, 1998).

19. For a discussion of the metaphor and its limits, see Didier Gascuel, *Pour une révolution dans la mer: De la surpêche à la résilience* (Arles: Actes Sud, 2019).

20. Callum Roberts, *The Unnatural History of the Sea: The Past and the Future of Humanity and Fishing* (London: Gaia, 2007).

21. The expression "tragedy of the commons" was coined in 1968 by Garrett Hardin to describe an economic problem similar to the bioeconomy of fisheries. Garrett Hardin, "The Tragedy of the Commons," *Science* 162, no. 3859 (1968): 1243–1248. Emphasizing the collective strategies required to overcome crises, Elinor Ostrom demonstrated the limits of Hardin's pessimism and suggested a brighter perspective on the management of the commons and socioecological governance. Elinor Ostrom, *Governing the Commons: The Evolution of Institutions for Collective Action* (Cambridge: Cambridge University Press, 2015). See also Gascuel, *Pour une révolution*, 440–456.

22. Rögnvaldur Hannesson, *The Privatization of the Oceans* (Cambridge, MA: MIT Press, 2006).

23. Allan Sekula, *Fish Story* (Düsseldorf: Richter, 1995). See also the documentary film by Allan Sekula and Noël Burch, *The Forgotten Space* (Amsterdam: Doc.Eye Film, 2010).

24. Lithograph by Charles Livingston Bull, United States Food Administration, ca. 1917–1918.

25. Howard Scott Gordon, "The Economic Theory of a Common-Property Resource: The Fishery," *Journal of Political Economy* 62, no. 2 (April 1954): 124–142.

26. Jeffrey A. Hutchings, "Collapse and Recovery of Marine Fishes," *Nature* 406 (2000): 882–885.

27. On ecological amnesia and the shifting baseline, see Roberts, *The Unnatural History of the Sea*; Daniel Pauly, "The Shifting Baseline Syndrome of Fisheries," in *Vanishing Fish: Shifting Baselines and the Future of Global Fisheries* (1995; Vancouver: Greystone Books, 2019), 94–98; Jeremy B. C. Jackson, Karen E. Alexander, and Enric Sala, eds., *Shifting Baselines: The Past and the Future of Ocean Fisheries* (Washington, DC: Island Press, 2011).

28. Jeffrey Bolster, *The Mortal Sea: Fishing the Atlantic in the Age of Sail* (Cambridge, MA: Harvard University Press, 2014).

Chapter 1

1. *Moving Outports*, British Movietone, AP Archive, BM83417, newsreel film, 1961.

2. George Withers, "Reconstituting Rural Communities and Economies: The Newfoundland Fisheries Household Resettlement Program, 1965–1970," PhD thesis, Memorial University of Newfoundland, 2016. A first Centralization Plan ran from 1954 onward, to be replaced with the more ambitious federal-provincial agreement

Newfoundland Fisheries Household Resettlement Program, which ran from April 1965 over a period of fifteen years.

3. Maureen Power, "Re-configurations: A Town in Newfoundland Grows When Houses Are Floated in from Far-Flung Outports," in *Journeys: How Traveling, Fruit, Ideas and Buildings Rearrange Our Environment*, ed. Giovanna Borasi (Montreal: Canadian Centre for Architecture/Actar, 2010), 102, 109.

4. Miriam Wright, *A Fishery for Modern Times: The State and the Industrialization of the Newfoundland Fishery, 1934–1968* (Don Mills, ON: Oxford University Press, 2001).

5. Built in Aberdeen by John Lewis & Sons for Fresh Frozen Foods Ltd., completed April 29, 1954. Overall length: 280 feet 6 inches or 85.5 meters; gross tonnage: 2,605 tons.

6. Wright, *A Fishery for Modern Times*, 106.

7. Trawling, which means dragging nets across the seabed, is one of the most ecologically devastating fishing techniques. In his history of ravaging ecological systems, Callum Roberts identifies the fourteenth-century invention of the beam trawl as "The First Trawling Revolution." Before the Great War, the use of steam power boosted trawling on an unprecedented scale: the impact of this was immense. After World War II, stern ramps made trawlers even more efficient, and the introduction of factory freezer trawlers with facilities to process, freeze, and store the catches on board allowed them to extend their working

periods away from port. On the increasing pressure on ecological systems, see Roberts, *The Unnatural History of the Sea*. On trawling, see Robinson, *Trawling*.

8. This process was not without controversy, as demonstrated by the "cod wars" between Iceland and the United Kingdom (the first between 1958 and 1961, the second between 1972 and 1973), which led to the establishment of a fifty-mile Icelandic Exclusive Economic Zone (EEZ). After World War II, several countries claimed sovereignty over the sea around their continental shelves. In 1958, under the auspices of the United Nations, treaties were signed defining strategies to secure "sea law." In 1982, the legal concept of a continental shelf (which does not necessarily coincide with the geological concept) and EEZs were finally established, though it was only in 1994 that the United Nations Convention on the Law of the Sea (UNCLOS) came into force. While traditionally maritime sovereignty stretched up to 3 nautical miles from the coast, UNCLOS extended this to 12 nautical miles, with EEZs being applicable for up to 200 nautical miles. On maritime jurisdiction, see Hannesson, *The Privatization of the Oceans*. For a Newfoundland perspective, see Wright, *A Fishery for Modern Times*, 128–140.

9. Once the international fleets were forced off the Grand Banks, the increasing pressure placed on the cod population by Canadian industrial fisheries depleted

it and led to the 1992 fishing moratorium. On the role of science in the disastrous depletion of the northern cod population, see Alan Christopher Finlayson, *Fishing for Truth: A Sociological Analysis of Northern Cod Stock Assessments from 1977–1990* (St. John's, NL: Institute of Social and Economic Research, Memorial University of Newfoundland, 1994).

10. The history of the resettlement programs is far from consensual. See also James Candow, "Books Reviews, Resettlement," *Newfoundland and Labrador Studies* 36, no. 1 (2021), 145–149. See also Isabelle Côté and Yolande Pottie-Sherman, eds., *Resettlement: Uprooting and Rebuilding Communities in Newfoundland and Labrador and Beyond* (St. John's, NL: ISER Books, 2020). For a classic visual assessment of the resulting landscape, see Scott Walden, *Places Lost: In Search of Newfoundland's Resettled Communities* (Toronto: Lynx Images, 2003). For a literary depiction of a twentieth-first century resettlement, see Michael Crummey, *Sweetland* (Toronto: Doubleday, 2014).

11. Jeff A. Webb, *Observing the Outports: Describing Newfoundland Culture, 1950–1980* (Toronto: University of Toronto Press, 2016).

12. Shannon Ryan, *Fish Out of Water: The Newfoundland Saltfish Trade 1814–1914* (St. John's, NL: Breakwater Books, 1986).

13. On the colonial cod fisheries in Newfoundland, see Peter Pope, *Fish into Wine: The Newfoundland Plantation in the Seventeenth Century* (Chapel Hill: University of

North Carolina Press, 2004). On nineteenth-century population shifts, see especially pp. 205–206.

14. Sean T. Cadigan, *Hope and Deception in Conception Bay: Merchant-Settler Relations in Newfoundland, 1785–1855* (Toronto: University of Toronto Press, 1995).

15. Biologists have related the "aerobic capacity of the mitochondria" to its "haemoglobin polymorphism" to realize that the genetic pool of different species populations varies according to the water's acidity, salinity, and temperature. Ole Brix, "The Physiology of Living in Water," in *Handbook of Fish Biology and Fisheries*, vol. 1, *Fish Biology*, ed. Paul J. B. Hart and John D. Reynolds (Oxford: Blackwell Science, 2004), 87.

16. Jeff A. Hutchings, "Life Histories of Fish," in Hart and Reynolds, *Handbook of Fish Biology*, vol. 1, 149–174.

17. Brix, "The Physiology of Living in Water," 89.

18. João Neves et al., "Population Structure of the European Sardine *Sardina pilchardus* from Atlantic and Mediterranean Waters Based on Otolith Shape Analysis," *Fisheries Research* 243 (2021): 1–10.

19. For a classic essay on upwelling, see David Cushing, "Upwelling and the Production of Fish," *Advances in Marine Biology* 9 (1971): 255–334. For a specific account of the phenomenon on the Portuguese coast, see F. Santos, M. Gómez-Gesteira, M. de Castro, and I. Álvarez, "Upwelling along the Western Coast of the Iberian Peninsula: Dependence of

Trends on Fitting Strategy," *Climate Research* 48, nos. 2/3 (2011): 213–218.

20. These paragraphs follow my paper "The Invention of Cod in Gafanha da Nazaré," originally published in *Spool* 8, no. 1 (2021): 113–138.

21. For a comprehensive history of cod fisheries, see Kurlansky, *Cod*.

22. On the Vikings in Newfoundland, see Jared Diamond, *Collapse: How Societies Choose to Fail or Survive* (2005; London: Penguin Books, 2011), 178–276.

23. Roberts, *The Unnatural History of the Sea*, 199–213.

24. Although the Beothuk avoided contact, Europeans contributed to the extinction of the Indigenous population by transforming their ecosystem, preying on the same species, and disrupting food chains, as well as introducing diseases. The last Beothuk died in 1829 of tuberculosis.

25. James Carscadden and Hjálmar Vilhjálmsson, "Capelin: What Are They Good For?," *ICES Journal of Marine Science* 59 (2002): 863–869; Alejandro D. Buren et al., "Bottom-Up Regulation of Capelin, a Keystone Forage Species," *PLOS One* 9, no. 2 (2014): 1–11; Laura M. Bliss et al., "Using Fisher's Knowledge to Determine the Spatial Extent of Deep-Water Spawning of Capelin (*Mallotus villosus*) in Newfoundland, Canada," *Frontiers in Marine Science* 9 (2023): 1–13.

26. On the importance of bait fisheries, with a focus on New England shores, see Bryan J. Payne, *Fishing a Borderless Sea: Environmental

Territorialism in the North Atlantic, 1818–1910* (East Lansing: Michigan State University Press, 2010).

27. For a definition of "regime shift," see chapter 3, "The Harbor and the Factory," 97–99. On the role of capelin for the recovery of Newfoundland cod populations see George Rose and Richard O'Driscoll, "Capelin Are Good for Cod: Can the Northern Stock Rebuild without Them?," *ICES Journal of Marine Science* 59 (2002): 1018–1026.

28. Moses Harvey, *Newfoundland as It Is in 1894: A Hand-Book and Tourists' Guide* (St. John's, NL: J. W. Wither, 1894), 40–41. The first road between St. John's and Portugal Cove was built in 1825. For a curious description of the drama of the railroad construction, see John Gimlette, *The Theatre of Fish: Travels through Newfoundland and Labrador* (London: Arrow Books, 2006), 133–135.

29. Such was the case for Boston and other growing commercial cities in New England. See Kurlansky, *Cod*, 62–106. Daniel Vickers, *Farmers and Fishermen: Two Centuries of Work in Essex County, Massachusetts, 1630–1850* (Chapel Hill: University of North Carolina Press, 1994).

30. This process was first highlighted by Harold Innis, *The Cod Fisheries: The History of an International Economy* (New Haven, CT: Yale University Press, 1940). Innis delves into the trade network sustained by cod fisheries from the sixteenth to the twentieth century, linking the human geographies of the Atlantic

to Newfoundland cod populations.

31. Claire-Desire Letourneur, ca. 1821, *Manuscript Atlas of the French Cod Fisheries of Newfoundland*, Memorial University of Newfoundland, Archives and Special Collections, Coll-477.

32. See the photographic album *Newfoundland Scenery*, presented to Joseph Laurence and attributed to Simeon H. Parsons (1844–1908), Memorial University of Newfoundland, Archives and Special Collections. See also Frances Rooney, *Working the Rock: Newfoundland and Labrador in the Photographs of Edith S. Watson, 1890–1930* (Portugal Cove-St. Philip's: Boulden Publications, 2017).

33. Today, sardines along the Portuguese coast are typically between six and seven years old.

34. Sue Shephard, *Pickled, Potted, and Canned: How the Art and Science of Food Preserving Changed the World* (2000; New York: Simon & Schuster, 2006). On the development of canning, see Rebeca Garcia and Jean Adrian, "Nicolas Appert: Inventor and Manufacturer," *Food Reviews International* 25, no. 2 (April 2009): 115–125.

35. Sardine fisheries were (and still are) a seasonal enterprise, active from early spring to late summer when the fish accumulate fat to produce the energy required for reproduction. Hence, sardine fishing became a supplementary occupation to local agricultural labor, consequently shifting the cultural axis of the region.

36. For a comprehensive description of Furadouro and its architecture, see Domingos Tavares, *Casas na duna: O Chalé do Matos e os palheiros do Furadouro* (Porto: Dafne Editora, 2018).

37. An equivalent rationale developed in Norway, where both sprats (*Sprattus*, which, like sardines, are forage fishes) and cod were studied on the basis of their different physiological qualities. Ellefsen and Lundevall, *North Atlantic Coast*.

38. Peter Sinclair, *A Question of Survival: The Fisheries and Newfoundland Society* (St. John's, NL: Institute of Social and Economic Research/ Memorial University Newfoundland, 1988); Lawrence F. Felt, "On the Backs of Fish: Newfoundland and Iceland's Experience with Fishery-Induced Capital Goods Production in the Twentieth Century," in Sinclair, *A Question of Survival*, 45–73. See also Jennifer M. Hubbard, *A Science on the Scales: The Rise of Canadian Atlantic Fisheries Biology, 1898–1939* (Toronto: University of Toronto Press, 2006), 127–128.

39. Ragnar Arnason, *The Icelandic Fisheries: Evolution and Management of a Fishing Industry* (Oxford: Blackwell Science/Fishing News Books, 1995). On the historical relation between urbanization and herring, see Lawrence C. Hamilton et al., "Sea Changes Ashore: The Ocean and Iceland's Herring Capital" in *Arctic* 57, no. 4 (2004): 325–335.

40. Between 1845 and 1884, when Newfoundland's population more than tripled and the proportion of fisherfolk kept stable at 80 percent. Sean T. Cadigan and Jeffrey A. Hutchings, "Nineteenth-Century Expansion of the Newfoundland Fishery for Atlantic Cod: An Exploration of Underlying Causes," in *The Exploited Seas: New Directions for Marine Environmental History*, ed. Poul Holm, Tim D. Smith, and David J. Starkey (St. John's, NL: International Maritime Economic History Association/Census of Marine Life, 2001), 41.

41. James E. Candow, "The Evolution and Impact of European Fishing Stations in the Northwest Atlantic," *Studia Atlantica* 3 (1999): 9–33.

42. Candow transcribes two detailed accounts of early fishing stations, one from 1671 by Nicolas Denys on the French stations, another by James Yong (1647–1721) on British outposts in 1663. Candow, "Evolution and Impact," 12–15.

43. In 1987, Mellin settled in Tilting to document its landscape and survey its constructions, building up a substantial oral history and countering primary sources with archival documents. Robert Mellin, *Tilting: House Launching, Slide Hauling, Potato Trenching, and Other Tales from a Newfoundland Fishing Village* (New York: Princeton Architectural Press, 2003). See also Robert Mellin, *Winter in Tilting: Slide Hauling in a Newfoundland Outport* (St. John's, NL: Pedlar Press, 2015).

44. The fragile architecture of fishing stations required continuous maintenance, if not reconstruction, each structure being reworked two or three times during its useful lifetime. When destroyed by storms or

the sea, most of its wood components could be retrieved from nearby shores to be recycled. This malleability and permanent reuse were reinforced by the spectacular moves that houses were often subjected to. Mellin, *Tilting*, 66.

45. Mellin, *Tilting*, 157–158.

46. Daniel Woodley Prowse, *A History of Newfoundland from the English, Colonial, and Foreign Records* (London: MacMillan, 1895), 366.

47. An example is the relative importance of immigration from Ireland, often credited as the source of Newfoundland's vernacular architecture. John J. Mannion, *Irish Settlements in Eastern Canada: A Study of Cultural Transfer and Adaptation* (Toronto: University of Toronto Press, 1974). For other insights on Newfoundland vernacular architecture, see Gerald L. Pocius, "Architecture on Newfoundland's Southern Shore: Diversity and the Emergence of New World Forms," *Bulletin of the Society for the Study of Architecture in Canada* 8, no. 2 (1983): 12–19. See also Rodrigue Girardin and Gerald Pocius, *Saint-Pierre et Miquelon, architecture et habitat* (Saint-Pierre and Miquelon: Éditions de L'Arche, 2006); Shane O'Dea, "Simplicity and Survival: Vernacular Response in Newfoundland Architecture," *Society for the Study of Architecture in Canada Bulletin* 8, no. 2 (1983): 4–11; Shane O'Dea, "The Tilt: Vertical-Log Construction in Newfoundland," *Perspectives in Vernacular Architecture* 1 (1982): 55–64.

48. Whereas British fisheries were mainly conducted from shore, prompting colonization, French fisheries were predominantly migratory and conducted on the offshore fishing banks, with the shore fishing stations serving as seasonal supports. There are several reasons why British settlements never got the needed traction to form a colony, a dynamic changed by the Napoleonic Wars, after which France retreated almost completely from fisheries, resuming only on bank fisheries. Newfoundland's population doubled roughly from 10,000 to 20,000 between 1793 and 1815; permission for Newfoundlanders to possess private property was granted by the British government in 1813, the first governor was appointed in 1818, and the first constitutional government in 1824 (Candow, "Evolution and Impact," 26–27).

49. "In 1763 the resident population was 7,000; in 1785 it had increased to 10,244; in 1804 it was found to be 20,380 and in 1834 it was 75,000." Harvey, *Newfoundland*, 20. Harvey's numbers do not match the more reliable numbers produced by Cadigan and Hutchings, "Nineteenth-Century Expansion," 41, which charted a growth from approximately 21,000 in 1845 to 72,000 in 1884.

50. Cadigan and Hutchings, "Nineteenth-Century Expansion," 38. See also R. A. Meyers, N. J. Barrowman, and J. A. Hutchings, "Inshore Exploitation of Newfoundland Atlantic Cod (*Gadus morhua*) since 1948 as Estimated from Mak-Recapture Data," *Canadian Journal of Fisheries and Aquatic Sciences* 54, suppl. 1 (1997): 224–s35.

51. Cadigan and Hutchings, "Nineteenth-Century Expansion," 39.

52. Jeffrey A. Hutchings and Ransom A. Myers, "What Can Be Learned from the Collapse of a Renewable Resource? Atlantic Cod, *Gadus morhua*, of Newfoundland and Labrador," *Canadian Journal of Fisheries and Aquatic Sciences* 51, no. 9 (September 1994): 2126–2146.

53. Cadigan and Hutchings, "Nineteenth-Century Expansion," 31.

54. Cadigan and Hutchings, "Nineteenth-Century Expansion," 40.

55. Cadigan and Hutchings, "Nineteenth-Century Expansion," 61.

56. Gordon, "The Economic Theory of a Common-Property Resource."

57. Later, in 1968, Garrett Hardin would coin the expression "the tragedy of the commons" while addressing a similar economic problematic to that examined by Scott Gordon. The topic will be discussed further in the chapter "The Lugger and the City."

58. A standard example is the importance of *garum* production during the Roman Empire, especially on the Tróia peninsula south of Lisbon. Sally Grainger, *The Story of Garum: Fermented Fish Sauce and Salted Fish in the Ancient World* (Abingdon: Routledge, 2020).

59. Inês Amorim, "A estrutura das 'artes novas' da costa de Aveiro, ao longo

da 2.ª metade do século XVIII: Mão-de-obra, divisão de trabalho, formas de propriedade e divisão do produto," in *Antropoloxía mariñeira*, ed. Francisco Calo Lourido (Santiago de Compostela: Consello da Cultura Galega, 1998), 159–182.

60. The new settlements were not intended to serve the needs of the existing population, who fished using earlier techniques.

61. I am grateful to Miguel Figueira, who taught me about the relationship between *arte xávega* and the surf. This topic was explored in the exhibition *Our Land Is the Sea: The Sensitive Construction of the Coastline* that we co-curated with an enthusiastic team for Garagem Sul at the Centro Cultural de Belém, Lisbon, 2020. See Miguel Figueira, *O mar é a nossa terra* (Porto: Pierrot-le-fou, 2020).

62. Octávio Lixa Filgueiras, "The *Xavega* Boat: A Case Study in the Integration of Archaeological and Ethnological Data," in *Sources and Techniques in Boat Archaeology*, ed. Sean McGrail (Greenwich: National Maritime Museum, 1977), 77–114.

63. Raquel Soeiro de Brito, *Palheiros de Mira: Formação e declínio de um aglomerado de pescadores* (Lisbon: Centro de Estudos Geográficos, 1960), 32–35.

64. Ernesto Veiga de Oliveira and Fernando Galhano, *Palheiros do Litoral Central Português* (Lisbon: Centro de Estudos de Etnologia Peninsular, 1964), 93–101. For a survey of

Portuguese building cultures, see *Arquitectura popular em Portugal* (Lisbon: Sindicato Nacional dos Arquitectos, 1961). Ethnographers distinguish the northern timber-built *palheiros* from their counterparts on the sandbars of the Algarve, built of straw in service of the local tuna fishery. This explains the use of the word *palheiro*— derived from *palha*, which means "straw"—to describe them. On these subtleties, see Ernesto Veiga de Oliveira, Fernando Galhano, and Benjamin Pereira, *Construções primitivas em Portugal* (1969; Lisbon: Dom Quixote, 1988), 189–248.

65. Oliveira and Galhano, *Palheiros*. Their study followed an earlier survey by Rocha Peixoto, "Habitação: Os Palheiros do Littoral," in *Portugália: Materiaes para o estudo do povo portuguez*, vol. 1 (1899–1903), 79–96, reprinted in Rocha Peixoto, *Etnografia Portuguesa (Obra Etnográfica Completa)* (Lisbon: Dom Quixote, 1990), 70–88. See also the case study on Soeiro de Brito, *Palheiros de Mira*. Apart from these sources, there is little documentation on the topic.

66. More than 300 *palheiros*, both houses and warehouses, were destroyed in each of these fires. Tavares, *Casas na duna*, 21–22, 45–48.

67. Amorim, "A estrutura das 'artes novas' da costa de Aveiro," 197; Tavares, *Casas na duna*, 32–33.

68. Charles le Goffic, "La crise sardinière," *Revue des deux mondes* (January 1907): 4–48; João Ferreira Dias and Patrice Guillotreau, "Fish Canning Industries of France

and Portugal: Life Histories," *Economia Global e Gestão* 10, no. 2 (2005): 61–79.

69. Tavares, *Casas na duna*, 39.

70. Although by increasing average landings and thus supply, the purse-seine fisheries had also contributed to a drop in the market value of sardines.

71. The *palheiros* did not survive much beyond the 1970s, but *xávegas* still operate from Espinho to Nazaré. Today's boats and nets are smaller, the oxen were replaced by motors in the 1980s, and the fresh fish caught are sold for local consumption, mainly to restaurants frequented by tourists.

Chapter 2

1. C. Scott Baker and Phillip J. Clapham, "Modelling the Past and Future of Whales and Whaling," *Trends in Ecology and Evolution* 19, no. 7 (July 2004): 365–371.

2. Francisco González de Canales, "Eladio and the Whale," *AA Files* 75 (2017): 152–161.

3. Randall Reeves and Tim Smith, "A Taxonomy of World Whaling," in *Whales, Whaling, and Ocean Ecosystems*, ed. James Estes et al. (Berkeley: University of California Press, 2006), 82–101.

4. Eric Jay Dolin, *Leviathan: The History of Whaling in America* (New York: W. W. Norton, 2007).

5. Herman Melville, *Moby-Dick; or, The Whale* (New York: Harper & Brothers, 1851).

6. Tim D. Smith et al., "Spatial and Seasonal Distribution of American Whaling and Whales in the

Age of Sail," *PLOS One* 7, no. 4 (April 2012): e34905.

7. Sophie Monsarrat et al., "A Spatially Explicit Estimate of the Prewhaling Abundance of the Endangered North Atlantic Right Whale," *Conservation Biology* 30, no. 4 (August 2016): 783–791.

8. Joshua Drew et al., "Collateral Damage to Marine and Terrestrial Ecosystems from Yankee Whaling in the 19th Century," *Ecology and Evolution* 6, no. 22 (November 2016): 8181–8192.

9. Nathaniel Philbrick, *Away Off Shore: Nantucket Island and Its People, 1602–1890* (London: Penguin, 2011).

10. Michael May, *Sconset House by House* (Nantucket, MA: Nantucket Preservation Trust, 2018).

11. Elizabeth A. Little, "The Indian Contribution to Along-Shore Whaling at Nantucket," *Nantucket Algonquian Studies*, vol. 8 (Nantucket, MA: Nantucket Historical Association, 1981).

12. Philbrick, *Away Off Shore*, 80.

13. J. Hector St. John [de Crèvecoeur], *Letters from an American Farmer* (Dublin: John Exshaw, 1782), 125. See also Nathaniel Philbrick, "The Nantucket Sequence in Crèvecoeur's Letters from an American Farmer," *New England Quarterly* 64, no. 3 (September 1991): 414–432. See also "'Every Wave Is a Fortune': Nantucket Island and the Making of an American Icon," *New England Quarterly* 66, no. 3 (September 1993): 434–447.

14. Equivalent lookouts were still used by Azorean shore fisheries in the middle of the twentieth century.

Strategically located upon the cliffs, they were not only important in observing the sea but also shaped a communication network that also functioned on land by linking the lookouts with boats and with fisherfolk dispersed in settlements on several islands (by means of fire, flags and sails, and later radio). On the Azorean fisheries, see Robert Clarke, *Open Boat Whaling in the Azores: The History and Present Methods of a Relic Industry*, Discovery Reports 26 (Cambridge: Cambridge University Press, 1954), 281–354, and Paulo Gouveia, *Arquitectura baleeira nos Açores* (Whaling architecture in the Azores) (Angra do Heroísmo: Gabinete de Emigração e Apoio às Comunidades Açorianas, 1995).

15. Henry Chandlee Forman, *Early Nantucket and Its Whale Houses* (1961; Nantucket, MA: Mill Hill Press, 1991), 31–43. Various sources reproduce a drawing in the first Proprietors' Book from 1776: Philbrick, *Away Off Shore*, 83; Nantucket Historical Association (NHA), Nantucket Registry of Deeds, 1775, SC740; May, *Sconset*, xii, fig. 2.

16. Alexander Starbuck, *The History of Nantucket* (Boston: C. E. Goodspeed, 1924), 356. Philbrick quotes Obed Macy's "Anecdotes," an unpublished NHA manuscript: "Copied from a book of Obed Macy's writing when he was an old man: 'If ever my History of Nantucket is republished in a second edition some of the following anecdotes may be found useful.'" Philbrick, *Away Off Shore*, 84. Partially

published in Starbuck, *History of Nantucket*.

17. Philbrick, *Away Off Shore*, 130.

18. Obed Macy, *The History of Nantucket, Being a Compendious Account of the First Settlement of the Island by the English, Together with the Rise and Progress of the Whale Fishery; and Other Historical Facts Relative to Said Island and Its Inhabitants* (Mansfield, MA: Macy & Pratt, 1835), 122, 212.

19. Philbrick, *Away Off Shore*, 108.

20. Obed Macy, "A Short Memorial of Richard Macy, Grandfather of Obed Macy," in *Anecdotes*, quoted in Philbrick, *Away Off Shore*, 106–107.

21. Macy, *The History of Nantucket*, 260–262.

22. Dolin, *Leviathan*, 207; Philbrick, *Away Off Shore*, 188–189.

23. Philbrick, *Away Off Shore*, 82; Little, "Indian Contribution to Along-Shore Whaling"; Daniel Vickers, "The First Whalemen of Nantucket," *William and Mary Quarterly* 40, no. 4 (October 1983): 560–583.

24. Charles Haskins Townsend, "The Distribution of Certain Whales as Shown by Logbook Records of American Whaleships," *Zoologica* 19, nos. 1–2 (April 1935). Townsend's work builds on the preceding series by Matthew Fontaine Maury, *Whale Chart of the World* (Washington, DC: United States Naval Observatory, 1852).

25. Smith et al., "Spatial and Seasonal Distribution." For an updated and accessible survey, see Judith Lund and Tim Smith, "American

Offshore Whaling Voyages," *Whaling History*, Mystic Seaport Museum and New Bedford Whaling Museum, https://whalinghistory.org/av.

26. Dolin, *Leviathan*, 72–74.

27. Hal Whitehead, "Sperm Whales in Ocean Ecosystems," in Estes et al., *Whales, Whaling, and Ocean Ecosystems*, 324–325.

28. Males "sometimes use areas less than 100 m deep." Whitehead, "Sperm Whales in Ocean Ecosystems," 325.

29. Macy, *The History of Nantucket*, 152.

30. For a description of the camels, see William C. Macy's addition ("Part Third") to Obed Macy, *History of Nantucket*, 2nd ed. (Mansfield, MA: Macy & Pratt, 1880), 286–287.

31. Randall R. Reeves, Jeffrey M. Breiwick, and Edward D. Mitchell, "History of Whaling and Estimated Kill of Right Whales, *Balaena glacialis*, in the Northeastern United States, 1620–1924," *Marine Fisheries Review* 61, no. 3 (1999): 1–36.

32. Phillip J. Clapham and Jason S. Link, "Whales, Whaling, and Ecosystems in the North Atlantic Ocean," in Estes et al., *Whales, Whaling, and Ocean Ecosystems*, 314–323.

33. Lance E. Davis, Robert E. Gallman, and Karin Gleiter, *In Pursuit of Leviathan: Technology, Institutions, Productivity, and Profits in American Whaling, 1816–1906* (Chicago: University of Chicago Press, 1997), 136. On the social behavior of sperm whales, see Peter B. Best, "Social Organization in Sperm Whales, *Physeter macrocephalus*," in *Behavior of Marine Animals*, ed. H. E. Winn et al. (New York:

Plenum Press, 1979), 227–289; Peter B. Best, "Sperm Whale Stock Assessments and the Relevance of Historical Whaling Records," *International Whaling Commission* 5 (1983): 41–55.

34. Davis, Gallman, and Gleiter, *In Pursuit of Leviathan*.

35. Davis, Gallman, and Gleiter, *In Pursuit of Leviathan*, 137–138.

36. If the chase was not targeting large bulls, a random distribution would result in 50 percent immature youngsters, 25 percent females, and 25 percent large males. Davis, Gallman, and Gleiter, *In Pursuit of Leviathan*, 137–138.

37. Elizabeth A. Josephson, Tim D. Smith, and Randall R. Reeves, "Depletion within a Decade: The American 19th-Century North Pacific Right Whale Fishery," in *Oceans Past: Management Insights from the History of Marine Animal Populations*, ed. David J. Starkey, Poul Holm, and Michaela Barnard (London: Earthscan, 2008), 133–147.

38. Josephson, Smith, and Reeves, "Depletion within a Decade," 142, 146.

39. Macy, *The History of Nantucket*, 217.

40. Macy, *The History of Nantucket*, 235.

41. Edouard A. Stackpole, "Nantucket Whale Oil and Lightning," *Historic Nantucket* 32, no. 4 (April 1985): 24–28.

42. Between 1815 and 1821, Nantucket's fleet recovered from twenty-five to seventy-eight ships. Dolin, *Leviathan*, 207n8; Macy, *The History of Nantucket*, 216.

43. Macy, "Part Third," 287–289. See also Philbrick, *Away Off Shore*, 226–229.

44. Paul Butler, *St John's: City of Fire* (Paradise, NL: Flanker Press, 2007).

45. On how natural resources and architectural design fueled the Chicago Fire, see William Cronon, *Nature's Metropolis: Chicago and the Great West* (New York: W. W. Norton, 1991), 179–180.

46. NHA MS37, The Great Nantucket Fire Collection, Box 3, Folder 19, "Report of the committee setting out the burnt districts streets to be widened," n.d.

47. Nathaniel Philbrick, *In the Heart of the Sea: The Tragedy of the Whaleship Essex* (2000; New York: Penguin, 2001), 221.

48. J. Congdon, *Map of New Bedford* (Boston: Pendleton, 1834).

49. Joseph L. McDevitt, *The House of Rotch: Massachusetts Whaling Merchants, 1734–1828* (New York: Garland, 1986).

50. Dolin, *Leviathan*, 213.

51. Joseph D. Thomas et al., *A Picture History of New Bedford*, vol. 1, *1602–1925* (New Bedford, MA: Spinner, 2013), 106–111.

52. Joe Silvia, "The Grand Designs of Russel Warren: New Bedford Architecture," *New Bedford Guide*, http://www.newbedfordguide.com; Thomas W. Puryear, *The Architecture of Russell Warren in the Coastal Towns of Southeastern New England* (North Dartmouth, MA: The Gallery, 1982); Arthur Channing Downs, *The Architecture and Life of the Hon. Thornton MacNess Niven (1806–1895), with Accounts of Architecture and Building Practices in Newburgh, Goshen, Monticello, and Riverhead, N.Y., and of Newly Discovered*

Architecture by Andrew J. Downing, A. J. Davis, Russell Warren, and Calvin Pollard (Goshen, NY: Orange County Arts Community of Museums & Galleries, 1972). On the city's cosmopolitanism, see John D. Kelly, "New Bedford, Capital of the 19th Century?," *Anthropological Quarterly* 90, no. 4 (Fall 2017): 1085–1126.

53. For the Egyptian mania of the 1850s and 1860s, see Stephanie Moser, *Owen Jones, Ancient Egypt and the Crystal Palace* (New Haven, CT: Yale University Press, 2012).

54. Leonard Bolles Ellis, *History of New Bedford and Its Vicinity 1602–1892* (Syracuse, NY: D. Mason, 1892), 495–502. The station was replaced in 1886.

55. Davis, Gallman, and Gleiter, *In Pursuit of Leviathan*, 342–380.

56. Davis, Gallman, and Gleiter, *In Pursuit of Leviathan*, 357.

57. New Bedford Whaling Museum, "New Bedford Cordage Company Records," 1839–1968, Mss 1.

58. Joseph D. Thomas et al., eds., *A Picture History of New Bedford*, vol. 1, 82–83. For a map of oil refineries operating between 1800 and 1860, see Marc Foster, "New Bedford: Whale Oil Refining Capital," *IA: The Journal of the Society for Industrial Archeology* 40, nos. 1–2 (2014): 54.

59. Davis, Gallman, and Gleiter, *In Pursuit of Leviathan*, 344–345.

60. Kingston Wm. Heath, "Whalers to Weavers: New Bedford's Urban Transformation and Contested Identities," *IA: The Journal of the Society for Industrial Archeology* 40, nos. 1–2 (2014): 7–32.

61. Seth H. Ingalls, "Map & Profile of Branch Rail-Road & Wharf in New Bedford," State Library of Massachusetts, Maps Collection.

62. Richard A. Voyer, Carol Pesch, Jonathan Garber, Jane Copeland, and Randy Comeleo, "New Bedford, Massachusetts: A Story of Urbanization and Ecological Connections," *Environmental History* 5, no. 3 (July 2000), 357. See also Carol Pesch et al., *Imprint of the Past: Ecological History of New Bedford Harbor* (Boston: US Environmental Protection Agency, 2001).

63. Kingston Wm. Heath, "Housing the Worker: The Anatomy of the New Bedford, Massachusetts, Three-Decker," *Perspectives in Vernacular Architecture* 10 (2005): 47–59, 359.

64. Joseph Samson, "A Description of Nantucket," *Port Folio* 5, no. 1 (January 1811): 36.

65. Dolin, *Leviathan*, 342–352; Elmo Paul Hohman, *The American Whaleman: A Study of Life and Labor in the Whaling Industry* (New York: Longmans, Green, 1928), 293–295. See also the bestseller by Peter Nichols, *Final Voyage: A Story of Arctic Disaster and One Fateful Whaling Season* (New York: G. P. Putnam's Sons, 2009).

66. Dolin, *Leviathan*, 343. Quoted in "Letter About the Arctic, no. II," in *Whalemen's Shipping List and Merchants' Transcript*, May 10, 1853.

67. Dolin, *Leviathan*, 347.

68. *Bird's Eye View of the Town of Nantucket, State of Massachusetts, Looking Southwest 1881*, Nantucket Historical Association, SC735; O. H. Bailey, *Bird's-Eye View of New Bedford, Mass.* (New Bedford, MA: Leonard B. Ellis, Fine Art Rooms, 1876).

Chapter 3

69. "La poissonnerie de Trouville," in *La construction moderne* 52, no. 22 (April 1937): 465–471; Cité de l'Architecture et du Patrimoine, Institut Français d'Architecture, Centre d'archives d'architecture du XXe siècle, fonds 163 IFA, Concours de la nouvelle poissonnerie de Trouville, 1935–1937, VINMA/35.7.

70. Jean-Claude Vigato, *L'architecture régionaliste: France 1890–1950* (Paris: Institut Français d'Architecture/ Norma, 1994), 199.

71. Pope, *Fish into Wine*.

72. Denis Binet, *Les pêches côtières de la baie du Mont-Saint-Michel à la baie de Bourgneu au XIXe siècle* (Plouzané: Ifremer, 1999); Xavier Dubois, *La révolution sardinière: Pêcheurs et conserveurs en Bretagne Sud au XIXe siècle* (Rennes: PUR, 2004). See also Constant Friconneau, *La saga de la sardine et du thon: Histoire de la pêche et de la conserve de Nantes aux côtes de Vendée* (Le Château d'Olonne: D'Orbestier, 1999).

73. For a good description of the techniques used, see *De la pêche de la sardine et des industries qui s'y rattachent, par un pêcheur* (Quimperlé: Imprimerie de C. Clairet, 1864).

74. Duhamel du Monceau, *Traité général des pesches et histoire des poissons qu'elles fournissent*, 2:432–436.

75. Dubois, *La révolution sardinière*, 27–28. The average size of such presses would be 70 square meters.

76. Pascal Brioist and Jean-Christophe Fichou, "La sardine à l'huile ou le premier aliment industriel: Nicolas Appert et Joseph Colin; Une filiation douteuse," *Annales de Bretagne et des Pays de l'Ouest* 119, no. 4 (2012): 69–80.

77. Shephard, *Pickled, Potted, and Canned*; Rebeca Garcia and Jean Adrian, "Nicolas Appert: Inventor and Manufacturer," *Food Reviews International* 25, no. 2 (April 2009): 115–125.

78. Caillo Jeune, *Recherches sur la pêche de la sardine en Bretagne et sur les industries qui s'y rattachent* (Nantes: Vincent Forest, 1855); Michel Mollat du Jourdin, *Histoire des pêches maritimes en France* (Toulouse: Privat, 1987).

79. Pascal Le Floc'h et al., "Identification des points de rupture dans la série longue dcs productions de sardine en France (1900–2017)," *Revue d'économie industrielle* 170 (2020): 48–78.

80. Brioist and Fichou, "La sardine à l'huile," 74; Binet, *Les pêches côtières*, 178.

81. Nathalie Meyer-Sable, "Spatialisation de l'habitat des marins-pêcheurs Etel, Morbihan, au xixème siècle," in *Environnements portuaires/Port Environments*, ed. Anne-Lise Piétri-Lévy, John Barzman, and Éric Barré (Mont-Sainte-Aignan: PURH, n.d.), 298–306.

82. Jean-Christophe Fichou, "L'entente des Travaux Maritimes et des conserveries de poisson en Bretagne méridionale (1850–1914)," in Piétri-Lévy, Barzman, and Barré, *Environnements portuaires*, 307–315. See also Dubois, *La révolution sardinière*, 38–42.

83. Dubois, *La révolution sardinière*, 75–77.

84. Dubois, *La révolution sardinière*, 111.

85. On the fish of the Gulf of Biscay surveyed in 1973 and 1976, see Jean-Claude Quéro, Jean Dardignac, and Jean-Jacques Vayne, *Les poissons du Golfe de Gascogne* (Plouzane: Ifremer/Muséum National d'Histoire Naturelle, 1989).

86. Florian Quemper, "Modélisation de la distribution spatiale de la sardine du Golfe de Gascogne (Sardina pilchardus) par intégration de données commerciales et scientifiques: Enjeux et limites," diploma thesis, Institut national supérieur des sciences agronomiques, de l'alimentation et de l'environnement, Rennes, 2020–2021.

87. Suzanne Arbault and Nicole Lacroix, "Oeufs et larves de clupeides et engraulides dans le golfe de Gascogne (1969–1973): Distribution des frayeres; Relations entre les facteurs du milieu et la reproduction," *Revue des Travaux de l'Institut des Pêches Maritimes* 41, no. 3 (September 1977): 227–254.

88. Suzanne Arbault and Nicole Lacroix, "Aires de ponte de la sardine, du sprat et de l'anchois dans le golfe de Gascogne et sur le plateau celtique: Résultats de six années d'étude," *Revue des Travaux de l'Institut des Pêches Maritimes* 35, no. 1 (March 1971): 50.

89. Connie Y. Chiang, *Shaping the Shoreline: Fisheries and Tourism on the Monterey Coast* (Seattle: University of Washington Press, 2008). See also John Radovich, "The Collapse of the California Sardine Fishery: What Have We Learned?," in *Resource Management and Environmental Uncertainty: Lessons from Coastal Upwelling Fisheries*, ed. Michael H. Glantz and J. Dana Thompson (1981; New York: Wiley, 1982), 107–136.

90. Alec D. MaCall, "The Sardine-Anchovy Puzzle," in Jackson, Alexander, and Sala, *Shifting Baselines*, 47–76.

91. Tsuyoshi Kawasaki, "Why Do Some Pelagic Fishes Have Wide Fluctuations in Their Numbers? Biological Basis of Fluctuation from the Viewpoint of Evolutionary Ecology," in *Proceedings of the Expert Consultation to Examine Changes in Abundance and Species Composition of Neritic Fishes Resources*, ed. G. D. Sharp and J. Csirke, FAO Fisheries Report 291 (1983), 1065–1080.

92. Dubois, *La révolution sardinière*, 23.

93. Jean-Christophe Fichou, "La crise sardinière de 1902–1913 au coeur des affrontements religieux en Bretagne," *Annales de Bretagne et des Pays de l'Ouest* 116, no. 4 (December 2009): 149–170.

94. Dubois, *La révolution sardinière*, 205–207; Amédée Odin, *Les besoins de l'industrie de la sardine* (Paris: Institut international de bibliographie scientifique, 1897).

95. According to one account, moving a figure of Saint Joseph to a remote cliff by the sea drove sardines away. Eugène Herpin, "Pourquoi les sardines s'éloignent des côtes de Bretagne," *Revue des traditions populaires* 3, no. 2 (February 1888): 98–99.

96. Dubois, *La révolution sardinière*, 273.

97. Dubois, *La révolution sardinière*, 270.

98. Dubois, *La révolution sardinière*, 274.

99. Dubois, *La révolution sardinière*, 254–255.

100. Chris Reid, "Evolution in the Fish Supply Chain," in *A History of the North Atlantic Fisheries*, vol. 2, *From the 1850s to the Early Twenty-First Century*, ed. David J. Starkey and Ingo Heidbrink (Bremen: H. M. Hauschild, 2012), 27–57.

101. Robinson, *Trawling*, 23–33.

102. See "The Short Blue Fleet," https://shortbluefleet .org.uk/.

103. On ice houses and freezing facilities, see following chapter 4, "The Salt and the Freezer."

104. Charles Knight, ed., *London* (London: Charles Knight, 1842), 4:208.

105. Priscilla Metcalf, *The Halls of the Fishmongers' Company: An Architectural History of a Riverside Site* (London: Phillimore, 1977).

106. George Dood, *The Food of London* (London: Longman, Brown, Green and Longmans, 1856), 345. Bunning's market hall was built by John Jay (1805–1872).

107. Nikolaus Pevsner, *A History of Building Types* (Princeton, NJ: Princeton University Press, 1976), 235–256.

108. Pevsner, *A History of Building Types*, 238–240.

109. Dood, *Food of London*, 345.

110. Charles Knight, ed., *London* (London: Charles Knight, 1843), 4:193–208.

111. Knight, *London*, 4:202.

112. James Bird, "Billingsgate: A Central Metropolitan Market," *Geographical Journal* 124, no. 4 (December 1958): 468. See also William John Passingham, *London's Markets: Their Origin and History* (London: Sampson Low, Marston, 1935).

113. Knight, *London*, 4:206.

114. Robinson, *Trawling*, 27.

115. Ice "is largely used, however, in packing the fish to be sent away by rail; and as fast as the trawlers come in and land their catches, the fish is sold, packed and forwarded by the next passenger train. Most of the Brixham fish is consigned at first to Bristol, but long before it arrives there, telegrams are sent on from Brixham to direct the sending of different quantities to London or other markets, according to the orders which have been received. There are, probably, few business transactions so generally conducted by telegraph, as the sale of fish." Edmund W. H. Holdsworth, *Sea Fisheries* (London: Edward Stanford, 1877), 122.

116. Construction begun in 1874, and the market opened on July 20, 1877.

117. Bunning's most acclaimed building was the Coal Exchange, inaugurated in 1849, with a prominent cast-iron dome.

118. Dood, *Food of London*, 347.

119. Henry Mayhew, *London Labour and the London of the Poor: The Condition and Earnings of Those That Will Work, Cannot Work, and Will Not Work* (1851; London: Charles Griffin and Company, 1864), 64, 173–176.

120. Knight, *London*, 4:202.

121. John K. Walton, *Fish & Chips and the British Working Class, 1870–1940* (London: Leicester University Press, 1992); Panikos Panayi, *Fish and Chips. A Takeaway History* (London: Reaktion Books, 2014).

122. Mayhew, *London Labour*, 174.

123. "The Insanitary Condition of Billingsgate," *British Medical Journal* 1, no. 1524 (March 15, 1890): 616.

124. Roberts, *The Unnatural History of the Sea*, 146.

125. Henri Verrière, "Les ports de pêche modernes," in *5eme Congrès national des travaux publics français* (Paris: Association française pour le développement des travaux publics, 1924).

126. For a survey of the port's history, see Gérard Le Bouëdec and Dominique Le Brigand, *Lorient Keroman: Du port de pêche à la cité du poisson* (Rennes: Marines Éditions, 2014); Olivier Busson, *Le port de pêche de Lorient-Keroman: Histoire du premier port de pêche francais, des origines à nos jours* (Paris: L'Harmattan, 2015). See also François Frey, ed., *60ème anniversaire du port de pêche de Lorient Kéroman: Exposition historique 26 juin–4 octobre 1987* (Lorient: Chambre de Commerce et d'Industrie du Morbihan, 1987).

127. Charles Robert-Muller, "Le nouveau port de pêche de Lorient. Chalutage et charbon," *Annales de Géographie* 36, no. 201 (1927): 201; Charles Robert-Muller, *Pêches et pêcheurs de la Bretagne atlantique* (Paris: A. Colin, 1944). See also Gérard Le Bouëdec, "Modèles de développement portuaire et urbain à Lorient," in Piétri-Lévy, Barzman, and Barré, *Environnements portuaires*, 146.

128. In the decade between 1928 and 1938, the number of trawlers increased significantly. In 1928, there were 59 (of which 55 were powered by steam and 4 by motor), and in 1938, there were 182 (54 steam and 128 motor). Robert-Muller, *Pêches et pêcheurs*, 216.

129. Robert-Muller, *Pêches et pêcheurs*, 241–242.

130. Robert-Muller, *Pêches et pêcheurs*, 240.

131. Le Bouëdec, "Modèles," 139–143.

132. Pascal Boisson, *Émile Marcesche, 1868–1939: Capitaine d'industrie à Lorient* (Lorient: Archives Municipales de Lorient, 2012).

133. Boisson, *Émile Marcesche*, 69–92.

134. Rapport sur le développement et l'avenir du chalutage à vapeur à Lorient, 12 novembre 1906. Quoted in Boisson, *Émile Marcesche*, 71.

135. Le Bouëdec and Le Brigand, *Lorient Keroman*, 23–25.

136. On the regime shift in Britanny, see Amédée Odin, *Les besoins de l'industrie de la sardine* (Paris: Institut International de Bibliographie Scientifique, 1897). On the Portuguese canning industry, see Charles Le Goffic, "La crise sardinière," *Revue des Deux Mondes* 37 (1907): 4–48; João Ferreira Dias and Patrice Guillotreau, "Fish Canning Industries of France and Portugal: Life Histories," *Economia Global e Gestão* 10, no. 2 (2005): 61–79.

137. Verrière, "Les ports de pêche modernes," 5.

138. A. Pousson and L. Vert, *La Rochelle: Ses industries, ses ports* (La Rochelle: Imprimerie de l'Ouest, 1928).

139. Mickaël Augeron and Jean-Louis Mahé, *Histoire de La Rochelle* (La Crèche: La Geste, 2012), 158–167.

140. On the La Rochelle fish market, see Yves Gaubert, Jean-Louis Mahé, and Henri Moulinier, eds., *La halle à marée d'hier à aujourd'hui* (La Crèche: La Geste, 2018).

141. Henri Verrière, quoted in Frey, *60ème anniversaire du port de pêche de Lorient Kéroman*, 17.

142. Marshall Berman, *All That Is Solid Melts into Air: The Experience of Modernity* (1982; London: Verso, 2010), 37–86.

143. Berman, *All That Is Solid*, 61.

144. This paragraph resumes a short essay I wrote on André Cepeda's photographs of contemporary Portuguese engineering works. André Tavares, "Oblique Angle," in André Cepeda, *Obras de Contrução em Portugal: Built Works in Portugal* (Lisbon: PTPC Plataforma Tecnológica Portuguesa da Construção, 2022), 34–39.

145. Maybe it is worth connecting this to the lucrative experience of Monterey's fish reduction companies.

146. Luchino Visconti, *La terra trema* (165 min., Italy, Arteas/Universalia, 1948). Another cinematic reference on boat ownership is Louis Malle, *Alamo Bay* (98 min., TriStar Pictures, 1981). Here, a young and entrepreneurial Vietnamese immigrant fights with a white supremacist community in Texas to secure the ownership of a shrimp boat.

147. Joseph Kerzoncuf, *La pêche maritime: Son évolution en France et à l'étranger* (Paris: A. Challamel, 1917), 3.

Chapter 4

1. On the bird's-eye view, see Jacques Gubler, "L'aérostation, prelude à l'aviation?," in *Motions, Émotions: Thèmes d'histoire et d'architecture* (Gollion: InFolio, 2003), 69–87. Originally published in *Matières* 2 (1998): 6–20.

2. John Summerson, *Georgian London* (1945; New Haven, CT: Yale University Press, 2003), 1.

3. Paul R. Josephson, "The Ocean's Hot Dog: The Development of the Fish Stick," in *Fish Sticks, Sports Bras, and Aluminum Cans: The Politics of Everyday Technologies* (Baltimore: John Hopkins University Press, 2015): 10–30.

4. Cronon, *Nature's Metropolis*, xvii.

5. Etienne Delaire and Pierre Teissier, "Horizons, chaînes et rivages frigorifiques en France, 1900–1930," *Cahiers François Viète* 3, no. 8 (2020).

6. Susanne Freidberg, *Fresh: A Perishable History* (Cambridge, MA: Belknap Press, 2010).

7. For a systematic survey of the history of refrigeration, see Roger Thévenot, *Essai pour une histoire du froid artificiel dans le monde* (Paris: Institut International du Froid, 1978). See also Jonathan Rees, *Refrigeration Nation: A History of Ice, Appliances, and Enterprise in America* (Baltimore: Johns Hopkins University Press, 2013).

8. The vegetable business is a side story in John Steinbeck's 1952 novel, *East of Eden*. Elia Kazan's film

adaptation of the book, released in 1955, features James Dean in action inside an icehouse, where the warm wooden colors of the building insulation contrast with the bluish reflections of the ice blocks. The venture ends in tragedy when the produce perishes because of inefficient rail transportation.

9. The pyramidal Glacière at the Désert de Retz in Chambourcy near Paris is an example of the long tradition of local icehouses. See Jacques Gubler, *Dear Signora Tosoni: Postcards to Casabella 1982–1995* (Milan: Skira, 2005), 21. Originally published as "Postcard 13," *Casabella* 492 (June 1983). See also Thomas Moore, *An Essay on the Most Eligible Construction of Ice-Houses: Also, a Description of the Newly Invented Machine Called the Refrigerator* (Baltimore: Bonsal & Niles, 1803).

10. "Arlington Icehouses Burned," *Boston Evening Transcript*, May 31, 1894, 5.

11. "Sky Lit Up," *Boston Daily Globe*, May 31, 1894, 1.

12. Oscar Edward Anderson Jr., *Refrigeration in America: A History of a New Technology and Its Impact* (1953; Princeton, NJ: Princeton University Press, 2016), 99.

13. "The Cold Storage Palace at the Columbian Fair and Its Destruction by Fire," *Scientific American* 69, no. 4 (July 1893): 52–53.

14. Hubert Howe Bancroft, *The Book of The Fair: An Historical and Descriptive Presentation of the World's Science, Art, and Industry, as Viewed through the Columbian Exposition at Chicago in 1893* (Chicago: Bancroft, 1895), 328, 339.

15. "In a Funeral Pyre: Firemen Cremated in the Cold-Storage Building," *Chicago Tribune*, July 11, 1893, 1–2.

16. Rees, *Refrigeration Nation*, 31–33.

17. Rees, *Refrigeration Nation*, 199.

18. Michael Osman, *Modernism's Visible Hand: Architecture and Regulation in America* (Minneapolis: University of Minnesota Press, 2018), 45–80.

19. Regardless of this failure, which outdid that of the World's Fair disaster, Chicago's other cold storages were developing regular business. See "Another Cold Storage Company," *Chicago Tribune*, September 25, 1892, 30. The item refers to a cold storage built on Clinton and Kinzie Streets; in 1971, a similar eighty-eight-year-old building was demolished at 350 North Dearborn Street.

20. "Boston's Cold Corner," *Ice and Refrigeration Illustrated* 9, no. 6 (December 1895): 375–395.

21. "Quincy Market Cold Storage Warehouse," *American Architect and Building News* 12 (August 26, 1882): 98.

22. The cross section is published in Osman, *Modernism's Visible Hand*, 64. Boston Public Library, William G. Preston Collection, vol. 22, no. 8.

23. "Boston's Cold Corner," 388.

24. "The New Quincy Market Cold Storage," *Cold Storage and Ice Trade Journal* 16, no. 2 (August 1906): 29–31.

25. "Quincy Market Cold-Storage Co.," *Boston Globe*, January 23, 1908, 7. By June 18, the newspaper registered the "proposed new building" to be executed "under the supervision of J. R. Worcester & Co.," *Boston Globe*, June 18, 1908, 3.

26. "T Wharf to Be Site for Storage Plant," *Boston Post*, August 2, 1916.

27. "Quincy Market Cold Storage and Warehouse Company," *Boston Globe*, December 23, 1915, 11.

28. "Old T Wharf in New Hands," *Boston Evening Globe*, August 1, 1916, 1–2.

29. "Soon T Wharf Will Be Deserted," *Boston Sunday Globe*, November 30, 1913, 14.

30. George Smith, Samuel Mansfield, and Heman Harding, *Thirty-Second Annual Report of the Board of Harbor and Land Commissioners for the Year 1910* (Boston: Wright & Potter, 1911), 6–10.

31. "Of Handsome Appearance," *Boston Globe*, February 5, 1914, 8.

32. "The Successor to Historic T Wharf," *Boston Evening News*, March 17, 1914, 6.

33. "Whole Village on Fish Pier," *Boston Sunday Globe*, May 31, 1914, 57.

34. "The Successor to Historic T Wharf," 6.

35. Reyner Banham, *A Concrete Atlantis: U.S. Industrial Building and European Modern Architecture* (Cambridge, MA: MIT Press, 1989).

36. The desolate area also meant accessibility difficulties for the fisherfolk and workers. As a newspaper recorded, workers "found walking a hard, long job"; there were relatively cheap boats providing access and expensive rides on motor buses. "Transportation to the Fish Pier," *Boston Globe*, May 28, 1914, 4.

37. "T Wharf Flag Hauled Down," *Boston Sunday Globe*, March 29, 1914, 9. See also "Glories of the T Wharf Gone," *Boston Sunday Globe*, August 6, 1916, 54.

38. "New Fish Pier in Operation," *Boston Evening Transcript*, March 30, 1914, 3. By May, mackerel was being landed in such quantities that its price was "getting down to a level where others beside the rich can afford them." "Fresh Mackerel Getting Cheaper," *Boston Globe*, May 7, 1914, 9.

39. "New Storage Plant at Fish Pier," *Boston Globe*, June 22, 1914, 4.

40. Charles H. Stevenson, *Preservation of Fishery Products for Food* (Washington, DC: Government Printing Office, 1899), 359. An innovative patent for refrigerated wagons was filed in 1867. J. B. Sutherland, "Improved Refrigerator-Car," Letters Patent no. 71 423, United States Patent Office, November 26, 1867.

41. Stevenson, *Preservation of Fishery Products for Food*, 387–388.

42. George F. Clementor, "Plan Showing Regulation of Fish Weirs in the Town of Truro," June 1923, Truro Historical Society.

43. Stevenson, *Preservation of Fishery Products for Food*, 388.

44. Irma Ruckstuhl, *Old Provincetown in Early Photographs* (New York: Dover, 1987), 4.

45. Bay State Fish Freezer, Historical Society of Old Yarmouth,

46. Stevenson, *Preservation of Fishery Products for Food*, 381.

47. Stevenson, *Preservation of Fishery Products for Food*, 383.

48. For a thoroughly researched and balanced biographical account, see Mark Kurlansky, *Birdseye: The Adventures of a Curious Man* (New York: Anchor Books, 2013). On Birdseye precedents, see pp. 137–140.

49. Kurlanksy, *Birdseye*, 41.

50. Clarence Birdseye, "Method of Preserving Piscatorial Products," *Letters Patent* no. 1,511,824, United States Patent Office, October 14, 1924.

51. Birdseye, "Method of Preserving Piscatorial Products," 1.

52. Birdseye, "Method of Preserving Piscatorial Products," 4.

53. Kurlansky, *Birdseye*, 154–155.

54. Clarence Birdseye, "Method of Preparing Food Products," *Letters Patent* no. 1,773,079, United States Patent Office, August 12, 1930.

55. Kurlansky, *Birdseye*, 166–174.

56. Kurlansky, *Birdseye*, 173.

57. Kurlansky, *Birdseye*, 155.

58. Fitz Henry Lane, *View of the Town of Gloucester*, 1836, colored lithograph on paper, Cape Ann Museum, no. 1988.36.10. For a depiction of the Fort peninsula, see Gloucester Harbor, 1852, inv. 38. In the 1893 Columbian Exposition in Chicago, the Gloucester pavilion at the Fisheries Exhibition displayed a diorama—designed by the land surveyor David W. Low (1833–1919)—which included a 1:24 scale model of Gloucester waterfront. This exquisite model of the fish drying racks is currently on display at Cape Ann Museum.

59. Daniel Vickers, *Farmers & Fishermen: Two Centuries of Work in Essex County, Massachusetts, 1630–1830* (Chapel Hill: University of North Carolina Press, 1994).

60. Bolster, *The Mortal Sea*, 66–67. See also William B. Leavenworth, "The Changing Landscape of Maritime Resources in Seventeenth-Century New England," *International Journal of Maritime History* 20, no. 1 (2008): 33–62.

61. Rees, *Refrigeration Nation*, 119–139.

62. Rees, *Refrigeration Nation*, 164.

63. Anderson, *Refrigeration in America*, 271n107.

64. Anderson, *Refrigeration in America*, 272.

65. Einar Richter Hansen, *From War to Peace: The Second World War at Nordkapp* (Honningsvåg: Nordkapplitteratur, 1994), 5.

66. Despina Stratigakos, *Hitler's Northern Utopia: Building the New Order in Occupied Norway* (Princeton, NJ: Princeton University Press, 2020).

67. Bjørn-Petter Finstad and Julia Lajus, "The Fisheries in Norwegian and Russian Waters, 1850–2010," in *A History of the North Atlantic Fisheries*, vol. 2, *From the 1850s to the Early Twentieth-First Century*, ed. David J. Starkey and Ingo Heidbrink (Bremen: H. M. Hauschild, 2012), 226–237.

68. Ole Sparenberg, "Frozen Fillets from the Far North: German Demand for Norwegian Fish," in *Industrial Collaboration in Nazi-Occupied Europe: Norway in Context*, ed. Hans Otto Frøland, Mats Ingulstad, and Jonas Scherner (London: Palgrave Macmillan, 2016), 63–85.

69. Ingo Heidbrink, "Creating a Demand: The Marketing Activities of the German Fishing Industry, c.1880–1990," in *The North Atlantic Fisheries: Supply, Marketing and Consumption, 1560–1990*, ed. David J. Starkey and James E. Candow (Hull: North Atlantic Fisheries History Association, 2006), 138–139. See also Sparenberg, "Frozen Fillets," 70.

70. Sparenberg, "Frozen Fillets," 69–70.

71. Tim Schröder, "Kraft durch Fischsahne," *Mare* 102 (February–March 2014): 80.

72. George Orwell, *Coming Up for Air* (1939; London: Penguin, 2020), 25–27.

73. Sparenberg, "Frozen Fillets," 74.

74. Anderson, *Refrigeration in America*, 204, 275.

75. See chapter 3, "The Factory and the Harbor." See also Etienne Delaire and Pierre Teissier, "Horizons, chaînes et rivages frigorifiques en France, 1900–1930," *Cahiers François Viète* 3, no. 8 (2020): 51–90.

76. Ingo Heidbrink, "From Sail to Factory Freezer: Patterns of Technological Change," in Starkey and Heidbrink, *A History of the North Atlantic Fisheries*, 63. The *Weser*, which was designed for fisheries in the Baltic, was part of another contemporary German experiment, which involved freezing capacity being provided by a side trawler: this served as a reference for Britain's postwar development of the factory fishing ship.

77. Lene Buhl-Mortensen, Hanne Hodnesdal, and Terje Thorsnes, eds., *New Knowledge from Mareano for Ecosystem-Based Management* (Oslo: Mareano, 2015).

78. Odd Nakken, "Past, Present and Future Exploitation and Management of Marine Resources in the Barents Sea and Adjacent Areas," *Fisheries Research* 37 (1998): 23–35.

79. Arvid Hylen, Odd Nakken, and Kjell Nedreaas, "Northeast Arctic Cod: Fisheries, Life History, Stock Fluctuations and Management," in *Norwegian Spring Spawning Herring and Northeast Arctic Cod: 100 Years of Research and Management*, ed. Odd Nakken (Bergen: Institute of Marine Research/Tapir Academic Press, 2008), 83–118.

80. Nils Kolle, "The Norwegian Coast—Nature's Offerings," in *Fish, Coast and Communities: A History of Norway*, ed. Nils Kolle et al. (Bergen: Fagbokforlaget, 2017), 55.

81. Atle Døssland, "Fisheries Vitalise the Coastal Communities, 1750–1880," in Kolle et al., *Fish, Coast and Communities*, 156–163.

82. Ellefsen and Lundevall, *North Atlantic Coast*, 70.

83. Ellefsen and Lundevall, *North Atlantic Coast*, 49n2.

84. Ellefsen and Lundevall, *North Atlantic Coast*. See also my review of the book in the *Journal of Architecture* 26, no. 4 (June 2021): 565–569, and a conversation with one of the authors, in Karl Otto Ellefsen, André Tavares, and Diego Inglez de Souza, *Notes on Codfish Architecture* (Lisbon: Centro Cultural de Belém, 2021).

85. Ellefsen and Lundevall, *North Atlantic Coast*, 53–54.

86. Ellefsen and Lundevall, *North Atlantic Coast*, 56.

87. Nils Kolle, "Between Tradition and Modernity, 1880–1945," in Kolle et al., *Fish, Coast and Communities*, 180.

88. Padmini Dalpadado and Bjarte Bogstad, "Diet of Juvenile Cod (age 0–2) in the Barents Sea in Relation to Food Availability and Cod Growth," *Polar Biology* 27 (2004): 140–154.

89. Kolle, "The Norwegian Coast—Nature's Offerings," 27–30. See also O. R. Godø, "Fluctuation in Stock Properties of North-East Arctic Cod Related to Long-Term Environmental Changes," *Fish and Fisheries* 4 (2003): 121–137.

90. Ellefsen and Lundevall, *North Atlantic Coast*, 164.

91. Bjørn-Petter Finstad, "Finotro: Statseid fiskeindustri i Finnmark og Nord-Troms—fra plan til avvikling," PhD diss., University of Tromsø (2005), 32.

92. Kolle, "Between Tradition and Modernity," 194–198.

93. Finstad, "Finotro," 44.

94. Finstad, "Finotro," 42.

95. Finstad, "Finotro," 34.

96. Norwegian Ministry of Trade, "Om landsplan for kjøleanlegg langs kysten og bevilgning til dens gjennemførelse," Handelsdepartementet, 1932. Appendix 3: Innstilling fra Kjøle- og Fryserikomiteen, November 6, 1931, 80. Quoted in Finstad, "Finotro," 36.

97. Ellefsen and Lundevall, *North Atlantic Coast*, 165–166.

98. Finstad, "Finotro," 45.

99. Svein Jentoft and Bjørn-Petter Finstad, "Building Fisheries Institutions through Collective Action in Norway,"

Maritime Studies 17 (2018): 13–25.

100. Sparenberg, "Frozen Fillets," 75.

101. Sparenberg, "Frozen Fillets," 75–76.

102. Finstad, "Finotro," 51–52.

103. Finstad, "Finotro," 52. In Hammerfest the plant had recourse to 300 Ukrainian women.

104. Bjørn-Petter Finstad, "The Norwegian Fishing Sector during the German Occupation: Continuity or Change?," in Frøland, Ingulstad, and Scherner, *Industrial Collaboration in Nazi-Occupied Europe*, 389–415.

105. Trond Dancke, *Opp av ruinene: Gjenreisningen av Finnmark 1945–1960* (Oslo: Gyldendal, 1986).

106. Dancke, *Opp av ruinene*, 9.

107. Ellefsen and Lundevall, *North Atlantic Coast*, 172–174.

108. On the professional politics of postwar Norwegian architecture, see Ingrid Dobloug Roede, "Reconstruction on Display: *Arkitektenes høstutstilling* 1947–1949 as Site for Disciplinary Formation," master's thesis, Massachusetts Institute of Technology (2019).

109. Ingebjørg Hage, "Reconstruction of North Norway after the Second World War: New Opportunities for Female Architects?," *Acta Borealia: A Nordic Journal of Circumpolar Societies* 22, no. 2 (2005): 99–127.

110. Dancke, *Opp av ruinene*, 213–227.

111. Hage, "Reconstruction of North Norway," 106, fig. 4.

112. William Nygaard, *Fra gjenreising til nyreising: Regionplanmøtet i Alta 12.–18. juli 1948* (Oslo: Tanum, 1950).

113. Dancke, *Opp av ruinene*, 146.

114. Finstad, "Finotro," 79–83.

115. Finstad, "Finotro," 90.

116. Hylen, Nakken, and Nedreaas, "Northeast Arctic Cod," 87. For a description of the daily routine of each of these techniques, see Ellefsen and Lundevall, *North Atlantic Coast*, 115–117.

117. British Intelligence Objectives Sub-Committee, *Certain Aspects of the German Fishing Industry*, BIOS Final Report, no. 493, London, H. M. Stationery Office, 1946.

118. Bjørn-Petter Finstad, "The Frozen Fillet: The Fish that Changed North Norway," *International Journal of Maritime History* 16, no. 1 (June 2004): 27–41.

119. Hans Tveitsme, "Kjøleteknikk i fiskeindustrien," in *Glimt fra norsk kjøleteknisk historie*, ed. Nils W. Pettersen-Hagh and Sæbjørn Røsvik (Oslo: Norsk Kjøleteknisk Forening, 1986).

120. Olav Notevarp, letter to the fisheries department, 31 May 1949. Quoted in Finstad, "Finotro," 102. Also quoted in Ellefsen and Lundvall, *North Atlantic Coast*, 177. Notevarp was a senior chemical engineer who held significant positions in the fisheries department before the war. See also Finstad, "Finotro," 61.

121. Finstad, "Finotro," 140, table 3.2. In 1948–1949, the three European countries combined received 64 percent of the exports, while the US represented only 3 percent of Frionor sales. See also Bjørn-Petter Finstad, "Modernizing the Fishing: Regional Fisheries Policy in Northern Norway, 1945–1970," in *North Atlantic Fisheries: Markets and Modernisation*, ed. Poul Holm and David J. Starkey (Esbjerg: Studia Atlantica, 1998), 179–190, esp. 184.

122. Terje Finstad, "Varme visjoner og frosne fremskritt: Om fryseteknologi i Norge, ca. 1920–1965," PhD diss., NTNU, Trondheim (2011).

123. Finstad, "Finotro," 157, table 4.9.

124. Alf R. Jacobsen, *Fra brent jord til Klondyke: Historien om Findus i Hammerfest og norsk fiskeripolitikks elendighet* (Oslo: Universitetsforlaget, 1996).

125. Findus stands for "fruit industry," FruktINDUStri, a name adopted when the original Swedish firm Skånska Frukt-vin-& Likörfabriken (Scandic Fruit, Wine, and Liqueur Factory) removed alcohol from its brand.

126. Finstad, "Finotro," 193, table 5.5.

127. Filmavisen Norkse Film AS, no. 26, June 25, 1953, 4:48 to 6:18.

128. Ellefsen and Lundvall, *North Atlantic Coast*, 160–213.

129. Kolle refers to Måstad in the Lofoten as "a vibrant fishery community" that "was completely deserted in the years after World War II." Kolle, "Between Tradition and Modernity," 193.

130. Dancke, *Opp av ruinene*, 148, fig. 28.

131. Nils Kolle and Pål Christensen, "Disputing Maritime Domains, 1910–2010," in Kolle et al., *Fish, Coast and Communities*, 234–259.

132. Finstad, "Modernizing the Fishing," 188.

133. Gunnars Sætersdal and Arvid Hylen, "The Decline of the Skrei Fisheries," *Fiskeridirektoratets Skrifter: Serie Havundersøkelser* 13, no. 7 (1964): 56.

134. Sætersdal and Hylen, "The Decline of the Skrei Fisheries," 62.

135. Sætersdal and Hylen, "The Decline of the Skrei Fisheries," 66, fig. 7.

136. Sætersdal and Hylen, "The Decline of the Skrei Fisheries," 66.

137. Godø, "Fluctuation in Stock Properties."

138. Godø experimented with cod equivalent methodologies to studies on the entanglement between fisheries and climate conducted for the Norwegian herring populations. Reidar Toresen and Ole Johan Østvedt, "Variation in Abundance of Norwegian Spring-Spawning Herring (*Clupea harengus*, Clupeidae) throughout the 20th Century and the Influence of Climatic Fluctuations," *Fish and Fisheries* 1, no. 3 (2000): 231–256.

Chapter 5

1. On cod and Portuguese nationalism, see Sally C. Cole, "Cod, God, Country and Family: The Portuguese Newfoundland Cod Fishery," *Maritime Studies* 3, no. 1 (1990): 1–29; see also José Manuel Sobral and Patrícia Rodrigues, "O 'fiel amigo': O bacalhau e a identidade portuguesa," *Etnográfica* 17, no. 3 (2013): 619–649.

2. The history of cod under the Portuguese dictatorial regime, the Estado Novo, was studied in detail by Álvaro Garrido, who published a variety of books, articles, and volumes of essays. His doctoral thesis is the most relevant document: Álvaro Garrido, *O Estado Novo e a Campanha do Bacalhau* (Lisbon: Círculo de Leitores, 2003). A major synthetic contribution, preceding Garrido's history was Mário Coutinho, *História da pesca do bacalhau: Por uma antropologia do "fiel amigo"* (Lisbon: Editorial Estampa, 1985).

3. In the online archive of the Portuguese television channel (RTP), it is possible to see the 1971 benediction: "Benção dos navios bacalhoeiros," *RTP Arquivos*, April 4, 1971, https://arquivos.rtp.pt/conteudos/bencao-dos-navios-bacalhoeiros. See also "O Chefe do Estado assistiu no alto da Encosta do Restelo à bênção dos navios bacalhoeiros," *Diário de Lisboa*, April 4, 1965, 1st edn., 1, 16.

4. On the parallels between Italian, German, and Portuguese fascisms, see Tiago Saraiva, *Fascist Pigs: Technoscientific Organisms and the History of Fascism* (Cambridge, MA: MIT Press, 2016). On the origins of the fleet blessing, see Pedro Theotónio Pereira, *Memórias: Postos em que servi e algumas recordações pessoais* (Lisbon: Verbo, 1972), 1:235–246; Álvaro Garrido, "O Estado Novo e a pesca do bacalhau: Encenação épica e representações ideológicas," *Oceanos* 45 (January–March 2001): 124–134.

5. Álvaro Garrido, "Os bacalhoeiros em revolta: A 'greve' de 1937," *Análise Social* 37, no. 165 (2003): 1191–1211.

6. Álvaro Garrido, *Henrique Tenreiro: Uma biografia política* (Lisbon: Círculo de Leitores, 2009), 137, 268–269.

7. *Diário de Lisboa*, April 4, 1965, 16.

8. *Diário de Lisboa*, April 4, 1965, 16. The reference to the almonds comes from *Jornal do Pescador* 27, no. 316 (May 1965): 15–16.

9. Scott Gordon authored the foundational essay describing such mechanisms: Gordon, "The Economic Theory of a Common-Property Resource."

10. The innovation is attributed to Captain Sabot. Étienne Bernet, *La grande pêche morutière: L'aventure des voiliers terre-neuviers Fécampois (1815–1931)* (Nolléval: Éditions l'écho des vagues, 2014), 15–17.

11. Bernet, *La grande pêche morutière*, 35–37.

12. Musée des Terre-Neuvas, ed., *Doris doris: Le doris hier et aujourd'hui, à Fécamp et dans le monde* (Fécamp: Musée des Terre-Neuvas Fécamp, 2002).

13. Winslow Homer, *The Fog Warning*, 1885, Museum of Fine Arts, Boston.

14. Francisco Esteves, *Pesca do bacalhau 1836–1856 (Companhia de Pescarias Lisbonense e Ericeira)* (Ericeira: author edition, 2019).

15. Baldaque da Silva, *Estado actual das pescas em Portugal compreendendo a pesca marítima*, 445.

16. Jacob Frederico Torlade Pereira d'Azambuja, *Memória sobre a pesca do bacalháo offerecida à Companhia de Pescarias Lisbonense* (Lisbon: Typographia de Desidério Marques Leão, 1835), 12.

17. António Vítor Nunes de Carvalho, "Os primeiros

passos na modernização da frota bacalhoeira portuguesa, 1935–1945: Aspectos da construção naval," *Oceanos* 45 (January–March 2001): 108–120.

18. Robinson, *Trawling*.

19. Bernet, *La grande pêche morutière*, 193.

20. Bolster, *The Mortal Sea*, 240–242.

21. Ricardo Lisboa da Graça Matias, "Os arrastões do bacalhau (1909–1993)," master's thesis, Faculdade de Letras da Universidade de Lisboa (2016), 121. The *Elite* had a gross register tonnage of 513 tons, about half of the 1,220-ton average of the trawlers registered in the 1946 census on Portuguese cod-fisheries ships. The weeklong Atlantic crossing and its high fuel costs explain the Portuguese resistance to investing in cod-trawling fisheries, a resistance only overcome when diesel engines and larger ships started to be adopted.

22. Grémio dos Armadores de Navios da Pesca do Bacalhau, *Frota de Pesca do Bacalhau. Álbum* (Lisbon: Abel de Oliveira, 1946). For luggers, the gross tonnage average was 6,300 tons, and for trawlers, 15,900 tons, whereas the total cargo hold was 21,123 tons, 69 percent of it in the luggers. For a history of Portuguese cod-fishing ships, see João David Batel Marques, *A pesca do bacalhau: História, gentes e navios*, 4 vols. (Viana do Castelo: Fundação Gil Eanes, 2018–2019). See also Nuno Valério, "Quanto vale o mar na economia portuguesa?," in *A economia marítima existe*,

ed. Álvaro Garrido (Lisbon: Âncora, 2006).

23. Garrido, *O Estado Novo*, 156–183. *Diário do Governo*, edict no. 23,968, June 5, 1934, superseded by edict no. 27,150, October 30, 1936.

24. Marques, *A pesca do bacalhau*, 4:136–148.

25. Minister of Commerce and Industry to the Chair of the CRCB board of directors, January 10, 1935, CDI, CRCB/515/EST3/PRAT60, sheet 432.

26. Memorandum to the Minister of Commerce and Industry, November 5, 1938 CDI, CRCB-515-EST-3/PRAT60, sheets 272–298.

27. *Diário do Governo*, edict no. 27,658, April 21, 1937. On the corporative organization of the economy, see n. 23 above.

28. Higino de Matos Queiroz to the Minister of Commerce and Industry, July 21, 1937, CDI, CRCB/515/EST3/PRAT60, sheet 368.

29. Higino de Matos Queiroz to the Minister of Commerce and Industry, July 21, 1937, sheet 368.

30. Higino de Matos Queiroz to the Minister of Commerce and Industry, July 7, 1937, sheet 369.

31. Higino de Matos Queiroz to the Minister of Commerce and Industry, July 7, 1937, sheet 369.

32. Quintals corresponded to different weights depending on location. Currently abbreviated as ctw, it is short for hundredweight, equal to 112 pounds in the United Kingdom and 100 pounds in the United States. For cod fisheries units, see Ryan, *Fish Out of Water*, xii.

33. Acácio Castel Branco, *Secadouros da Praça de Aveiro*, June 1946. CDI, CRCB/EST11/PRAT20/Ui.14, p. 7. In 1947, a plan for Aveiro's harbor quotes and corrects this number, identifying between 1.47 and 1.77 square meters per quintal. Junta Autónoma da Ria e Barra de Aveiro, *Plano de Arranjo e Expansão do Porto Bacalhoeiro de Aveiro* (June 1947), 2.

34. On the relationship between different cod sizes and weights and their provenance (Greenland, Labrador, Grand Banks), as well as the relative weights of their parts, see Mário Ruivo, "Factores de conversão e índices de aproveitamento do Bacalhau (Gadus callarias L.): Investigações Portuguesas na área da ICNAF; Campanhas de 1954–55," *Boletim da Pesca* 11, no. 53 (1957).

35. Castel Branco, *Secadouros da Praça de Aveiro*, 6.

36. The racks were usually 1.5 meters wide with joists spaced at between 5 and 6 meters; the spelter cables that functioned as support for cod were raised 80 centimeters from the ground.

37. Maria Lamas, *As mulheres do meu país* (1948–1950; Lisbon: Caminho, 2002), 201–219.

38. CRCB, *Monografia dos secadouros nacionais*, 1957–1958, CDI, CRCB/EST11/PRAT20/Ui.12 a Ui.14.

39. In 1957, of the twenty-nine registered dryers, only ten possessed refrigerated warehouses, almost all built in the late 1940s.

40. Only five of the twenty-nine companies possessed artificial dryers.

41. In 2022, this complex, which had hardly been used since the early 1990s, showed signs of aging but was still more or less intact. Early projects and surveys are archived in CDI, CRCB/EST11/PRAT20/Ui.112, as well as in the private archive of PGP.

42. The company started its operations in the Azores Islands, after a merger between two branches of the Bensaúde family. Esteves, *Pesca do bacalhau 1836–1856*, 147. A 1940s advertisement claimed it was incorporated in 1891.

43. For a precise inventory of locations and their characteristics, see Tavares and Souza, *Arquitectura do bacalhau*, 68–69.

44. André Tavares, "The Invention of Cod in Gafanha da Nazaré," *Spool* 8, no. 1 (2021): 113–138; André Tavares, *Um retrato marítimo de Ílhavo: Ler a história de uma paisagem através do seu património marítimo, militar e industrial/A Maritime Portrait of Ílhavo: Reading the History of a Landscape through Its Maritime, Military and Industrial Heritage* (Guimarães: Lab2PT, 2021).

45. Saul António Gomes, ed., *Ílhavo: Terra milenar* (Ílhavo: Câmara Municipal de Ílhavo, 2017).

46. Moses Bensabat Amzalak, *A pesca do bacalhau* (Lisbon: Museu Comercial, 1923).

47. Two drying facilities in Porto had no boats registered in their name. In the 1957 inventory, twelve ships owned by the same company are associated with the main dryers in Vila Nova de Gaia and Alcochete, as well as in São Jacinto, despite the latter having been abandoned in 1952.

48. For the drawing, see António Craveiro Lopes, "Melhoramento da Ria de Aveiro, Rectificação da Margem da Gafanha," May 17, 1925, drawing, archive Associação dos Portos de Aveiro. Landings in Aveiro were registered from 1903 onward: see Manuel Ferreira Rodrigues, "O regresso à Terra Nova dos bacalhaus de navios armados em Aveiro e Ílhavo," *Oceanos* 45 (January–March 2001): 81ff.

49. Castel Branco, *Secadouros da Praça de Aveiro*, 1.

50. Castel Branco, *Secadouros da Praça de Aveiro*, 2.

51. Castel Branco, *Secadouros da Praça de Aveiro*, 9.

52. Castel Branco, *Secadouros da Praça de Aveiro*, 9.

53. Castel Branco, *Secadouros da Praça de Aveiro*, 10. He cites the example of a warehouse whose construction was embargoed by the port authority. In the 1957 monograph, these warehouses are marked as having been built in 1951, suggesting that the conflict had been solved in favor of the fisheries administration.

54. Garrido, *Henrique Tenreiro*, 311.

55. Júlio Ferreira David, *Relatório da Comissão de Estudos, no estrangeiro, de assuntos referentes à pesca, indústria e comércio de bacalhau*, September 1935, AR-AHP, PT-AHP/AN/CIOC/S11/SS8/Ui.1 (AN Cx. 58 no. 3).

56. Ferreira David, *Relatório da Comissão de Estudos*.

57. See chapter 4, "The Salt and the Freezer," 169–170.

58. André Tavares, *Duas obras de Januário Godinho em Ovar* (Porto: Dafne Editora, 2012), 34–37.

59. Nuno Paulo Soares Ferreira, "Entreposto Frigorífico do Peixe de Massarelos: Um dos ícones da arquitectura modernista portuense," master's thesis, Faculdade de Letras da Universidade do Porto (2010), 71.

60. OPCA had recently commenced its activities, combining civil engineering with electrotechnics and mechanical engineering. One of the four founding partners was the engineer Manuel Godinho (1898–1970), the elder brother of Januário, who by then had his office in a room of the construction company. Luís Lousada Soares, *Artes e letras na tradição das gentes da casa* (Lisbon: OPCA, 1992).

61. Manuel de Melo Cabral Vaz Guedes de Bacelar, *Entreposto do Peixe e Frigorífico* (Porto: Tipografia Universal, 1941), 3.

62. Bacelar, *Entreposto do Peixe e Frigorífico*, 17–18.

63. Bacelar, *Entreposto do Peixe e Frigorífico*, 26.

64. A. Coelho de Oliveira, "Os Serviços de Salubridade e Abastecimento. I. Relatório da actividade no ano de 1950," *Civitas* 1, nos. 2–3 (1945): 333–373.

65. *Diário do Governo*, edict no. 27 150, October 30, 1936, 1360.

66. Deolinda Folgado, "A caixa do frio artificial: A conformação de um lugar na Lisboa dos anos 40," in *Museu do Oriente: De armazém frigorífico a espaço museológico*

(Lisbon: Fundação Oriente, 2008), 48–61. See also João Paulo Martins, "João Simões (1908–1995), arquitecto: Armazéns frigoríficos e muito mais," in *Museu do Oriente*, 6–22.

67. In Salazar's archive, there is a set of 1931 correspondence to the head of state in which the trawler fishing companies beg for equivalent attention to be given to sardine and cod fisheries. However, such attention was not granted until the late 1950s, and both long-distance lugger cod fisheries and local purse-seine sardine fisheries were subsidized. ANTT, AOS/CO/MA-1, sheets 236–287.

68. "Memória descritiva," in *Concurso para a construção dos armazéns frigoríficos para a conservação de bacalhau sêco e frutas, a edificar em Lisboa, no terrapleno norte-nascente da Doca de Alcântara* (Lisbon: Editorial Império, 1939), 26–27.

69. The Ministry of Commerce and Industry planned to erect a different building, contiguous to the cod cold storage, to cater to fruit storage. Although it remained unbuilt, a full project was developed by the same engineer, Yglesias d'Oliveira. CDI, CRCB/EST11/PRAT19/Ui.01, vol. 2, March 1938.

70. Rui Neves Pereira, "Os armazéns frigoríficos da CRCB," *Panorama* 3, no. 18 (1943): 5–7.

71. CRCB, "Produção e importação: Estatística" (1947), 51 sheets. CDI, CRCB/EST11/PRAT19/Ui.07.

72. "Localização e capacidades dos armazéns frigoríficos destinados a bacalhau: estudo preliminar" (1941), CDI, CRCB/EST11/PRAT19/Ui.01, vol. 5, sheet 4.

73. "Intoxicação com bacalhau nos refeitórios dos serviços de Assistência Social da Legião Portuguesa," 1952, Arquivo Nacional Torre do Tombo, AOS/CO/PC-42. Salazar's correspondence indicates that major incidents of food poisoning caused by cod happened in 1949, 1951, and 1952, notably at Porto's civil penitentiary, and that "one of these cases of food poisoning was suffered by 800 inmates, some of whom have died as a consequence."

74. Mário Moutinho, *História da pesca do bacalhau: por uma antropologia do "fiel amigo"* (Lisbon: Estampa, 1985), 174–176.

75. Moutinho, *História da pesca do bacalhau*, 171.

76. Augusto Rocha Borges, "A furunculose dos pescadores. Possível 'fonte nocente' de toxi-infecções alimentares atribuídas ao peixe," in *IV Congresso Nacional de Pesca* (Lisbon: Gabinete de Estudos das Pescas, 1955), 241–242.

77. *Hum açoutado pelo Bacalhao ajoelhado ao rabo deste seu algoz, para que elle naõ açoute mais, ou O filho da Terra Nova naturalizado na Terra Farta* [One Battered by the Cod, Kneeling at the Tail of His Tormentor So That He Won't Batter Me Any More, or The Son of Newfoundland Naturalized in the Land of Plenty] (Porto: Viuva Alvarez Ribeiro & Filhos, 1826).

78. *Sentença de proscripção, que contra Dom Bacalhao alcançou Dona Sardinha, no tribunal dos cabazeiros, dentro na Torre de Pão, erecta na Cotovia* (Lisbon: Tipografia Régia, 1816). See also *O bacalhau justificado, ou Conversação do Futre Bacalhau com Dona Carne* (Porto, 1824).

79. *O adeos do Bacalhau, na presente quaresma de 1825* (Porto: Imprensa do Gandra, [1825]), 3.

80. *O adeos do Bacalhau*, 3–4.

81. Pope, *Fish into Wine*.

82. Ryan, *Fish Out of Water*, 148–169.

83. Ryan, *Fish Out of Water*, 148n1.

84. António Baldaque da Silva, *Estado actual das pescas em Portugal comprehendendo a pesca marítima, fluvial e lacustre em todo o continente do reino, referido ao anno de 1886* (Lisbon: Ministério da Marinha e Ultramar, 1891), 180.

85. Francisco Soares Franco, *História resumida da Companhia de Pescarias Lisbonense* (Lisbon: Typographia do Gratis, 1840).

86. Franco, *História resumida da Companhia de Pescarias Lisbonense*, 13–14.

87. This refers to the drying racks visible in several photographs of Portões de Santa Clara on São Miguel Island. See João Gomes Vieira, *O homem e o mar: Os açorianos e a Pesca Longínqua nos bancos da Terra Nova e Gronelândia* (Lisbon: Intermezzo-Audiovisuais, 2004), 82–83. Comprising several layers of intricate networks, Azores and Portuguese cod production had a complex relationship, with the Azorean family Bensaúde having a presence in the business of the important company Parceria Geral de Pescarias.

88. António Baldaque da Silva, *Restauração do poder marítimo de Portugal* (Lisbon: Livraria de António Maria Pereira, 1894), 16.

89. Baldaque da Silva, *Restauração do poder marítimo de Portugal*, 7.

90. Baldaque da Silva was explicit: the goal was to "replace the imports of foreign cod with the national production of the same fisheries; because by doing so, it is possible to apply thousands of *reis* [currency] for the benefit of Portugal's commerce and industry ... and train an excellent school of sailors for our own navigation purposes ... while also providing, during peacetime, a reserve army in case of war." Baldaque da Silva, *Restauração do poder marítimo de Portugal*, 27, 37.

91. Moses Bensabat Amzalak, *A pesca do bacalhau* (Lisbon: Museu Comercial, 1923).

92. Act of June 12, 1901. Amzalak, *A pesca do bacalhau*, 24.

93. Coutinho, *História da pesca do bacalhau*, 41. "Composição da Frota Bacalhoeira em 1917," quoted in Carvalho Brandão, *A pesca do bacalhau* (Lisbon: Pedro Bordalo Pinheiro/Livraria Profissional, 1917), 8–9. See also João Carlos de Oliveira Leone, *Inquérito à pesca do bacalhau: Feito por ordem do Conselho Central* (Lisbon: Liga Naval Portugueza, 1903).

94. Garrido, *O Estado Novo*, 47, 71–72.

95. "Rumo à Terra Nova," *Illustração Portugueza* 3 (May 20, 1907): 611–616; "A pesca do bacalhau," *Illustração Portugueza* 230 (July 18, 1910): 77–82; A. Mesquita de Figueiredo, "A pesca do bacalhau," *Illustração Portugueza* 368 (March 10, 1913): 306–309.

96. In 1926, he lasted two weeks as minister of finance; he was reappointed in 1928 with extended powers before being appointed president of the Government Council in 1932, with a new constitution in 1933 granting him increased executive powers. He remained in position until he literally fell off his chair after a stroke and was replaced in power in 1968, after forty years as head of state.

97. António Oliveira Salazar, "Alguns aspectos da crise das subsistências," in *António de Oliveira Salazar: O ágio do ouro e outros textos económicos (1916–1918)*, ed. Nuno Valério (Lisbon: Banco de Portugal, 1997), 223–275.

98. António Oliveira Salazar, "Sobre a indústria das conservas de peixe (Relatório de uma visita aos centros conserveiros)," in *Discursos e notas políticas, 1943 a 1950* (Coimbra: Coimbra Editora, 1951), 699–714.

99. Salazar, "Alguns aspectos da crise das subsistências," 238.

100. *Localisação e capacidade dos armazéns frigoríficos destinados a bacalhau: Estudo preliminar*, CDI, CRCB/EST11/PRAT19/Ui.01, vol. 5, no. O-555, no. R-5351, sheets 1–9.

101. Garrido, *O Estado Novo*, 306–318, esp. 307, 310.

102. George Simenon, *The Grand Banks Café* (London: Penguin, 2014). Originally published as *Au rendez-vous des Terre-Neuvas* (Paris: Fayard, 1931).

103. Bernet, *La grande pêche morutière*, 239–241.

104. Rudyard Kipling, *Captains Courageous: A Story of the Grand Banks* (London: Macmillan, 1897); translated into Portuguese by António Sérgio as *Lobos do mar* (Lisbon: Editorial Progresso, [193?]).

105. Rolf Lessenich, "The Mission of Empire: Romantic Humanitarianism in Rudyard Kipling's Earlier Work," *Studia Neophilologica* 65, no. 2 (1993): 197–211.

106. Álvaro Garrido, introduction to Alan Villiers, *A campanha do Argus: Uma viagem na pesca do bacalhau* (Lisbon: Cavalo de Ferro, 2005), 7–35. The first edition was published in English as *The Quest of the Schooner Argus* (London: Hodder & Stoughton, 1951), and the Portuguese translation by José da Natividade Gaspar appeared in the same year as *A campanha do Argus: Uma viagem aos bancos da Terra Nova e à Gronelândia* (Lisbon: Clássica, 1951).

107. The film *The Bankers: The Voyage of the Schooner Argus* was screened on March 15, 1951, in Lisbon's geographical society. In 1968, the National Geographic Society directed *The Lonely Dorymen: Portugal's Men of the Sea*, filmed on board the lugger *José Alberto*. Another 1966 documentary, *The White Ship*, was produced by the National Film Board of Canada and directed by Hector Lemieux. In 2018, an episode of *Land & Sea*, from the Canadian broadcaster CBC, recovered a 1967 work directed by John O'Brien,

Portuguese Fishermen on the Grand Banks of Newfoundland.

108. Villiers, The Quest of the Schooner Argus, 232–233.

109. Anita Conti, Racleurs d'océans (1953; Paris: Petite Biblio Payot, 2017); Deep Sea Saga, trans. Lynton Hudson (London: William Kimber, 1955). On Conti's photographs, see Laurent Girault-Conti, Anita Conti: Les Terre-Neuvas (Paris: Chêne, 2004).

110. On August 9, on the Fiskaerness bank, 40 kilometers off the coast of Greenland, Conti's trawler met a group of Portuguese dories. Conti, Racleurs d'océans, 174. Villiers narrates the encounter with Spanish trawlers on pages 63–64.

111. Describing the captains, Conti uses the same epic arguments, the hierarchy and authoritative discipline described by Kipling. "Each of these men … had been shaped, since childhood, by courage and pride. If their temperaments are different, there is a common trait uniting them: the ambition to succeed." Conti, Racleurs d'océans, 190.

112. Conti, Racleurs d'océans, 207.

113. Conti, Racleurs d'océans, 198–199.

114. Rachel Carson, The Sea Around Us (1951; London: Unicorn, 2014); translated into French as Cette mer qui nous entoure (Paris: Delamain et Boutelleau, 1952).

115. Helen Rozwadowski, Vast Expanses: A History of the Oceans (London: Reaktion Books, 2018), 199–202. Trained as a biologist, Carson published her first book,

Under the Sea Wind, in 1941. After the success of The Sea Around Us, she published The Edge of the Sea in 1955, in which she describes the distinct Atlantic marine habitats of coral and sandy and rocky shores off the United States. For a detailed biography, see Linda Lear, Rachel Carson: Witness for Nature (Boston: Mariner Books, 2009).

116. Rachel Carson, "The Silent Spring of Rachel Carson," CBS Reports, April 3, 1963, televised appearance transcribed in Lear, Rachel Carson, 450.

117. Rachel Carson, speech at National Book Award, January 29, 1952, quoted in Lear, Rachel Carson, 219.

118. See chapter 1, "The Cove and the Surf," 39–41.

119. Nuno Costa Silva e Rosário Vieira, "Os diários de bordo," Argos, Revista do Museu Marítimo de Ílhavo 4 (October 2016): 140–143.

120. Built in 1872, Labrador was a wooden-hulled lugger belonging to the company Casa Bensaúde & C.ª, Lda., which was involved in the cod fisheries from 1885 until 1904. The existing logbook refers to the 1894 campaign. Arquivo de Marinha, PT/ BCM-AH/0010/0044/1182.

121. Elite logbook, 1909– 1910, Arquivo de Marinha, PT/ BCM-AH/0628, 6/III/7/3.

122. For a map of these movements, see Tavares and Souza, Arquitectura do bacalhau, 56–57.

123. On the use of photography to document fishing, see Loren McClenachan, "Documenting Loss of Large Trophy Fish from the Florida Keys with

Historical Photographs," Conservation Biology 3, no. 23 (June 2009): 636–643.

124. On the measurements and relative sizes of living cod and their provenance (Greenland, Labrador, Grand Banks) for Portuguese fisheries in the 1950s, see Mário Ruivo, "Factores de conversão e índices de aproveitamento do Bacalhau (Gadus callarias L.): Investigações Portuguesas na área da ICNAF; Campanhas de 1954–55," Boletim da Pesca 11, no. 53 (1957); "Contribuição para o estudo das relações entre os comprimentos do bacalhau (Gadus callarias l.) inteiro, escalado e 'verde', pescado na Terra-Nova," Boletim da Pesca 11, no. 55 (1957): 3–10.

125. A popular example of such frequency is the 1884 advertisement for the cod liver oil Scott's Emulsion, which depicts a fisher carrying on his back a specimen weighing over 70 kilos "taken from life on the coast of Norway." The illustration of a cod, identified by its skin pattern and the three back fins, was based on a photograph by Marcus Selmer of a Bergen fisher carrying a large salmon on his back. The photograph, taken in 1872, was published as number 101 in the collection "Norwegian National Costumes." See online at http://www .digitaltmuseum.no.

126. John D. Harbron, "The Soviet's Floating City in Our Atlantic Waters," Maclean's, June 30, 1962. On the collapse of Newfoundland cod populations, see Finlayson, Fishing for Truth.

Epilogue

1. Roberts, *The Unnatural History of the Sea*; Bolster, *The Mortal Sea*.

2. Sigfried Giedion, *Mechanization Takes Command: A Contribution to Anonymous History* (1948; New York: Oxford University Press, 1970). See the chapter "Mechanization and Death: Meat," 209–245.

3. Cronon, *Nature's Metropolis*, 207–59. The classical depiction of Chicago slaughterhouses from a social angle is the novel by Upton Sinclair, *The Jungle* (1906; London: Penguin, 1985).

4. Jane Hutton, *Reciprocal Landscapes: Stories of Material Movements* (London: Routledge, 2020); Kiel Moe, *Unless: The Seagram Building Construction Ecosystem* (New York: Actar Publishers, 2020); Kiel Moe, *Empire, State & Building* (New York: Actar Publishers, 2017).

5. Nancy Couling and Carola Hein, eds., *The Urbanisation of the Sea: From Concepts and Analysis to Design* (Rotterdam: nai010 Publishers, 2020).

6. Stefanie Hessler, ed., *Prospecting Ocean* (Cambridge, MA: MIT Press, 2019); Stefanie Hessler, ed., *Tidalectics: Imagining an Oceanic Worldview Through Art and Science* (Cambridge, MA: MIT Press, 2018). See also MAP Office, *Our Ocean Guide* (Venice: Lightbox, 2017).

7. Mario Carpo, *The Alphabet and the Algorithm* (Cambridge, MA: MIT Press, 2011).

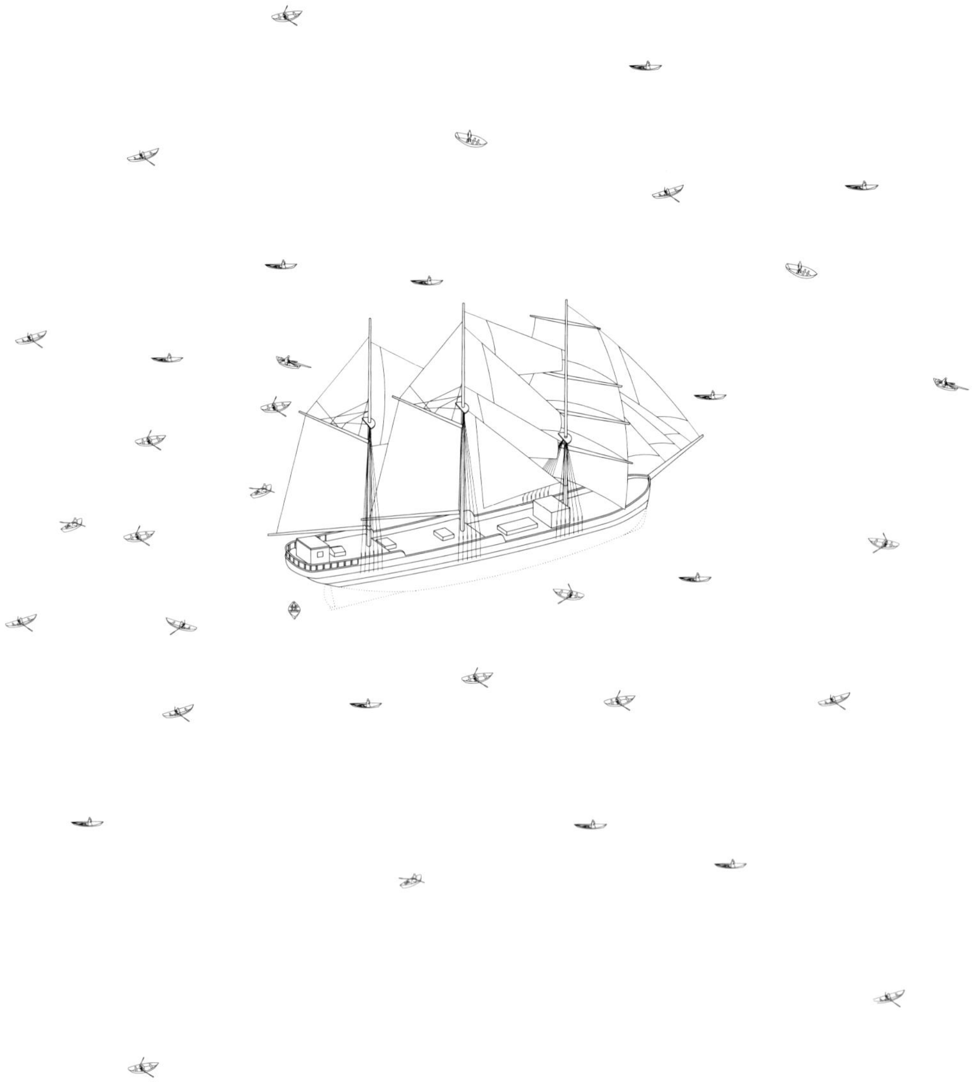

Published with the support of the Graham Foundation for Advanced Studies in the Fine Arts, Chicago, and the Faculty of Architecture of the University of Porto.

Graham Foundation

The *Fishing Architecture* research project was funded by the European Union (ERC, Fish-A, 101044244). The views and opinions expressed are, however, those of the authors only and do not necessarily reflect those of the European Union or the European Research Council. Neither the European Union nor the granting authority can be held responsible for them.

European Research Council
Established by the European Commission

The MIT Press would like to thank the anonymous peer reviewers who provided comments on drafts of this book. The generous work of academic experts is essential for establishing the authority and quality of our publications. We acknowledge with gratitude the contributions of these otherwise uncredited readers.

This book was set in Haultin by the MIT Press. Printed and bound in Canada.

Library of Congress Cataloging-in-Publication Data is available.

ISBN: 978-0-262-04910-8

10   9   8   7   6   5   4   3   2   1